Annette Holler

New Metrics for Value-Based Management

GABLER RESEARCH

Unternehmensführung & Controlling

Herausgegeben von
Universitätsprofessor Dr. Dr. habil. Wolfgang Becker,
Otto-Friedrich-Universität Bamberg
und Universitätsprofessor Dr. Dr. h.c. Jürgen Weber,
WHU – Otto Beisheim School of Management, Vallendar

Die Schriftenreihe präsentiert Ergebnisse der betriebswirtschaftlichen Forschung im Themenfeld Unternehmensführung und Controlling. Die Reihe dient der Weiterentwicklung eines ganzheitlich geprägten Management-Denkens, in dem das Controlling als übergreifende Koordinationsfunktion einen für die Theorie und Praxis der Führung zentralen Stellenwert einnimmt.

Annette Holler

New Metrics for Value-Based Management

Enhancement of Performance Measurement
and Empirical Evidence on Value-Relevance

With a foreword by Professor Dr. Dirk Schiereck

GABLER

RESEARCH

Bibliographic information published by the Deutsche Nationalbibliothek
The Deutsche Nationalbibliothek lists this publication in the Deutsche Nationalbibliografie;
detailed bibliographic data are available in the Internet at http://dnb.d-nb.de.

Dissertation European Business School, Oestrich-Winkel, 2009

D 1540

1st Edition 2009

Editorial Office: Claudia Jeske | Anita Wilke

Gabler is part of the specialist publishing group Springer Science+Business Media.
www.gabler.de

Umschlaggestaltung: KünkelLopka Medienentwicklung, Heidelberg
Printed on acid-free paper
Printed in Germany

ISBN 978-3-8349-1869-7

For Raphaela, Giulia, and Mauro

Foreword

Still today, the vast majority of the German DAX 30 corporations are explicitly comitted to the shareholder value idea. However, the stock return to shareholders is an inappropriate measure for internal performance management, so that companies revert to so-called Value-Based Management Systems. These systems derive accounting-based performance measures, which shall quantify whether and how much shareholder value has been generated by the management over a specific period (overall and by business unit). Data applied for calculating the performance measure has to be publicly available and, preferably, also separately disclosed, to allow capital market participants to project results of the performance measure on the stock price.

Various metrics (net earnings, residual income, EVA, REVA), which also demonstrated some correlation with stock returns in empirical studies, have been implemented for performance measurement in the past, but these measures still show methodological deficits and also the correlation with the stock performance can still be increased considerably. Also in this respect, the study of Annette Holler makes a substantial contribution, demonstrating that two new metrics (CRI, CEVA) show superior results both methodologically and in the empirical test. To date, such an examination of various yet existing and newly developed performance measures has not been known to me. Hence, the dissertation of Ms. Holler extends, in an amazing way, preceding academic research. In several also explorative analyses, a detailed picture of the information content of diverse measures of Value-Based Management is documented. Therewith, an objective state of knowledge is achieved, based on that well-founded recommendations for the design of the performance measurement and management system can be derived, which, in turn, should lead to a further increase in the effectiveness of corporate management.

Ms. Holler optimally fulfills the independently defined goals in her dissertation. The work entails numerous interesting results and is written in a way that it will surely be a pleasure for the reader to read to the end. I wish that this work will be paid the due attention it deserves.

Professor Dr. Dirk Schiereck
Chair for Corporate Finance
Faculty of Law and Economics
Technische Universität Darmstadt

Die ganz überwiegende Mehrzahl der deutschen DAX 30-Konzerne bekennt sich heute (immer noch) ganz explizit zum Shareholder Value-Gedanken. Als direkte Steuerungs-größe innerhalb eines Unternehmens ist die Aktienrendite allerdings ungeeignet, so dass auf sogenannte Value-Based Management Systeme zurückgegriffen wird. Diese Systeme leiten aus Rechnungslegungsdaten Steuerungsgrößen ab, die erfassen sollen, ob und wie viel Wert in der jeweils zurückliegenden Periode vom Management (insge-samt und heruntergebrochen auf Business Units) für die Aktionäre erwirtschaftet wurde. Damit von den Kapitalmarktteilnehmern ein Transfer von Ergebnissen der Steuerungsgröße auf den Aktienkurs vorgenommen werden kann, müssen die Daten zur Berechnung der Steuerungsgröße öffentlich verfügbar und möglichst auch noch separat ausgewiesen sein.

Verschiedene Maßzahlen (Net Earnings, Residual Income, EVA, REVA) wurden in der Vergangenheit als Steuerungsgrößen implementiert, zeigten auch in empirischen Studien gewisse Korrelationen zur Aktienrendite, weisen aber immer noch methodische Defizite auf, und auch die Korrelationen zur Aktienperformance sind noch deutlich steigerungsfähig. Auch hier setzt die Arbeit von Frau Holler an, um zu zeigen, dass zwei neue Maßzahlen (CRI, CEVA) sowohl methodisch als auch im empirischen Test überlegene Ergebnisse zeigen. Eine derartige Untersuchung ver-schiedener länger existierender und neu entwickelter Steuerungsgrößen ist mir bis dato nicht bekannt. Die Dissertationsschrift von Frau Holler betritt somit (in erstaunlicher Weise) wissenschaftliches Neuland. In mehreren, auch explorativ angelegten Untersuchungen wird ein detailliertes Bild des Aussagegehalts verschiedener Steuerungsgrößen des Value-Based Managements dokumentiert. Damit wird ein objektiver Kenntnisstand erreicht, auf dessen Basis sich fundierte Handlungs-empfehlungen für die Gestaltung der Performanceermittlung und Steuerungsmodelle ableiten lassen, die wiederum zu einer weiteren Effektivitätssteigerung der Unternehmenssteuerung führen sollten.

Frau Holler kann die selbst gesetzten Ziele in ihrer Dissertationsschrift bestens er-füllen. Die Arbeit enthält zahlreiche spannende Resultate und ist so geschrieben, dass es den Lesern sicherlich Freude machen wird, sie bis zum Ende zu lesen. Ich wünsche der Arbeit eine ihr gebührende weite Verbreitung.

Professor Dr. Dirk Schiereck
Lehrstuhl für Unternehmensfinanzierung
Fachbereich Rechts- und Wirtschaftswissenschaften
Technische Universität Darmstadt

Acknowledgments

This dissertation summarizes research I conducted during my doctoral studies at the chair for banking and finance of the EUROPEAN BUSINESS SCHOOL (ebs) at Oestrich-Winkel. Many people contributed to making this PhD study a valuable and successful experience. Without their support, I could neither have reached the heights nor explored the depths.

Above all, I am in full debt to my PhD advisor Professor Dr. Dirk Schiereck, who transferred by now from the ebs to the TU Darmstadt, for providing his committed supervision and expertise. He managed to focus my ideas, direct my enthusiasm, provide invaluable ideas, and to motivate me. I also wish to thank Professor Dr. Michael Henke from the EBS Oestrich-Winkel for volunteering as a second advisor and always providing constructive comments and valuable suggestions.

Moreover, other persons specifically contributed to my work: First of all, I very deeply appreciate the efforts of Mauro Risi, CEO Agip Deutschland and Benelux, who provided the initial ideas for this research, gave his time to read an early version of this work, and always understood to challenge and enrich my work. I also want to thank Professor Sanford J. Leeds from the University of Texas at Austin, with whom I verified the relevance of the research topic, for sharing his knowledge and offering me valuable contacts to discuss my research plans. I thank Felix Zeidler and Mike Stenglein for assisting me at all times with data requests. I very much acknowledge the support on the selection and implementation of statistical methods that I received from Dr. Wolfgang Rauch from the University of Frankfurt. At the same time, I would like to thank Alexander Schmidt for developing customized software that was a great help in performing the statistical analysis.

Additionally, I thank my employer Bain & Company for giving me valuable support and time off to conduct this study and providing me a stimulating and effective background. I want to thank Britta Heinrich for helping me to collect literature and beyond that all my colleagues for supporting my research intentions and numerous valuable discussions, in particular, Carsten Prussog, Christian Kinder, Arndt Kaminski, Pascal Schweizer, Joachim Krebs, and Heike Moses.

Last but not least, I would like to thank all my family and friends; particularly, Natalie, Susi, and Josef for their patience and persistence with regard to our friendship, Raphaela and Alex for their confidence and encouragement, my parents Anneliese and Konrad for their support, and Mauro, the love of my life, for his dedication.

Annette Holler
Endowed Chair for Banking and Finance
EUROPEAN BUSINESS SCHOOL (ebs)
International University Schloß Reichartshausen

List of Contents

3   Development of New Value-Based Metrics 75

4   Applied Methods for the Empirical Research 107

List of Figures

List of Tables

Abbreviations

AccAdj	Accounting adjustments
Accr	Accruals
ADA	Accumulated depreciation and amortization
AR	(Market-adjusted) abnormal stock returns
ATInt	After-tax interest
AV	Added value
BCG	Boston Consulting Group
BV	Book value of net assets
CapChg	Capital charge
CAPM	Capital asset pricing model
CE^{BI}	Cash earnings before interest
CEVA	Cash economic value added
CIP	Construction in progress
CFO	Cash flow from operations
CFROI	Cash flow return on investment
COPAT	Cash operating profits after taxes
CRI	Cash residual income
CROCE	Cash return on capital employed
CV	Corporate value
CVA	Cash value added
DA	Depreciation and amortization
DCF	Discounted cash flow
DD	Double declining balance depreciation method
Dep	(Accounting) depreciation
DepAdj[1]	Depreciation adjustment (including goodwill amortization)
DepAdj[2]	Depreciation adjustment (excluding goodwill amortization)
EBEI	Earnings before extraordinary items
ED	Economic depreication
EM	Economic margin
EOY	End of year balance
EP	Economic profit
ESOP	Employee stock ownership plan
EVA	Economic value added
FCF	Free cash flow
FGV	Future growth value
GA	Book value of gross assets
GDA	Gross book value of depreciable assets
GI	(Inflation-adjusted) gross investments

GIC	Gross invested capital
IC	Invested capital
IFRS	International financial reporting standards
IntExp	Interest expense
IRR	Internal rate of return
LIFO	Last-in-First-out
MV	Market value of the firm
MVA	Market value added
MVE	Market value of common equity
NA	Book value of net assets
NDA	Non-depreciable assets
NE^{BI}	Net earnings before interest
NI	Net investments
NIBCLs	Non-interest bearing current liabilities
NOPAT	Net operating profits after taxes
NOPLAT	Net operating profits less adjusted taxes
OCF	Operating cash flows
OGCF	Operating gross cash flows
OPAT	Operating profits after taxes
PI	Performance improvement
R&D	Reasearch and development
REVA	Refined economic value added
RI	Residual income
ROCE	Return on capital employed
ROGA	Return on gross assets
ROIC	Return on invested capital
RONA	Return on net assets
RR	(Unadjusted) raw stock returns
S/L	Straight-line depreciation method
SoY	Sum-of-the-years-digits depreciation method
SS	Stern Stewart & Company
SVA	Shareholder value added
TBR	Total business return
TCO	Total cost of ownership
TSR	Total shareholder return
US-GAAP	United States generally accepted accounting principles
VBM	Value-based management
WACC	Weighted average cost of capital
WAI	Wealth added index

Symbols

β	Beta factor of the share used within the CAPM
β_i	Regression coefficient
C	Incremental unique information component
D/E	Dept-to-equity ratio
δ	Constant parameter from a discrete linear stochastic process
r	Risk-adjusted discount rate
E	Future performance expectation
ε	Random disturbance term
FE_X	Forecast error for a performance measures X
M	Arithmetic mean
Φ	Autoregressive parameter from a discrete linear stochastic process
P	Stock price
R	Stock returns
r	Risk-adjusted rate
r_d	Cost of debt
r_e	Cost of equity
r_f	Risk-free rate of return
r_i	Stock return of firm i
r_m	Market-wide rate of return
r_S	Pearson product-moment correlation coefficient
SD	Standard deviation
t	Tax rate
T	Total useful life of an investment
T^*	Remaining useful life of an investment
w_d	Capital weight of debt capital
w_e	Capital weight of equity capital
X	Accounting-based measure of firm performance

1 Introduction

Substantial pressure to maximize shareholder wealth compels firms to adopt a value-based management system. As centerpiece of such a system, a value-based performance measure quantifies periodic value addition for shareholders. Prevalent value-based measures imply distortions from accounting depreciation and empirical evidence on prevailing measures is somewhat biased. Consequently, this study aims to develop depreciation-adjusted performance measures and to provide empirical evidence on prevailing and newly introduced measures. Specifically, this study concludes examining the relationship between bottom-line accounting measures of corporate performance with stock prices. However, in a broader sense, it also relates to strategic management and key drivers and dynamics of capital markets.

In the following, the underlying rationales are first presented and, afterwards, the objectives and the structure of the present study.

1.1 Motivation

1.1.1 Inadequacy of Performance Measures

This study refers to the basic concept of various value-based management systems, i.e., to consider maximization of shareholder value as main (when not even the only) overall corporate objective. Value-based management systems gained predominance in the 1990s and still prevail today. [1] At the core of an integrated value-based management system, periodical performance measures, based on financial statutory accounts or interim results, quantify management's additions to shareholder wealth and constitute the main reference for addressing decision making processes in all main financial, operating and general management activities, such as strategic planning, capital budgeting, acquisition pricing, performance measurement, internal communication, investor relations, and management compensation. [2]

External accountability of a value-based performance measure is essential. Only measures based on official financial data, publicly accessible, and on objective and

---

[1]   See section 2.1.1.
[2]   See Chew, Stern, and Stewart (1996) and O'Byrne and Young (2001, p. 18).

consistent methodologies can be used effectively by and communicated to the financial community, shareholders, and other stakeholders, to monitor management performance, assess investment and evaluate corporate value creation. At a first glance, stock prices, earnings, residual income, and economic value added, may be valid measures of the overall corporate performance. However, all these measures involve some problems and deficiencies for properly representing and quantifying shareholder value creation.

The change in stock price is the most obvious measure of added shareholder wealth. However, such market-derived measure is primarily driven by economy- and industry-wide factors beyond the control of management and other factors unrelated to the operating performance of the firm, such as noise trading and portfolio insurance.[3] Consequently, market-based stock returns constitute only indirect and often inappropriate value-based performance measures.

Among accounting measures, management normally considers net earnings as key performance indicator, responding to investors, who are primarily concerned about long-term earnings and use earnings as main input for valuation models.[4] Focus on earnings contradicts valuation theory, which assumes that expected future free cash flows determine stock price, and, thus, shareholder value.[5] Nonetheless, earnings prevail even after the introduction of the statement of cash flows in 1987, since the investment community continues to believe that earnings are a better indicator of a firm's present and future ability to generate free cash flows than cash flow information.[6] However, net earnings do not accurately measure changes in shareholder wealth: Positive earnings may imply positive or negative addition to shareholders' net worth, depending on the amount of equity costs.

Accordingly, 'Residual Income' (RI), defined as net earnings after the cost of debt and equity capital, represents an alternative accounting estimate, overcoming some defect of net earnings. Since RI includes charges for the cost of equity, realizing that equity capital has to be remunerated, it is a superior method compared to net earnings in describing changes in shareholder wealth. For centuries, economists used RI to describe the value creation of the firm as excess return after return expectations of investors given the systematic risk class of the firm's debt and equity capital.[7] However, RI was not considered until the twentieth century.[8] Further, only few large firms adopted RI for internal performance management or external reporting before

---

[3]   See Jensen and Murphy (1990) and Milbourn (1996).

[4]   See Foster (1986) and Lev (1989) for the predominance of earnings and the price-earnings ratio.

[5]   See section 4.2.2.

[6]   Introduction of the cash flow statement following FASB Statements of Standards No. 95, 1987; further, FASB's Statement of Financial Accounting Concepts No. 1, 1978, underlines the prevalence of earnings.

[7]   E.g., Hamilton (1777); equivalently, Marshall (1890, p. 142): *"What remains of his profits after deducting interest on his capital at the current rate may be called his earnings of undertaking or management."*

[8]   See Canning (1929), Edey (1957), Edwards and Bell (1961), Kay (1976), Peasnell (1981, 1982), and Preinreich (1936, 1937, 1938).

the emergence of value-based management, such as General Motors in the 1920s and General Electric in the 1950s.[9] However, also RI does not measure shareholder value creation accurately, since it is affected by several non-cash accruals and other accounting principles.

Consequently, several variants of RI, adjusting for such distortions, have been developed by business consultancies.[10] Among those, 'Economic Value Added' (EVA), developed by Stern Stewart & Company, prevails. The measure has been first mentioned in 1989.[11] By 1998, EVA has yet become a leading management tool.[12] Its popularity is attributable to a supporting press coverage, enthusiastic corporate sponsors, and bold claims by Stern Stewart & Company, as follows: First, extensive press coverage created broad awareness of the EVA concept in the business and investment community. E.g., Fortune magazine and the Journal of Applied Corporate Finance praised EVA as outstanding measure and published annual EVA league tables.[13] Second, well-known firms, which adopted EVA and publicly commented on their satisfaction, built further trust in EVA; moreover, managers commonly reported that EVA adoption motivated them to optimize capital utilization and, consequently, increase shareholder value.[14] In addition, major financial management firms and investment banks such as Calpers, Oppenheimer, Credit Suisse First Boston, and Goldman Sachs reported to use EVA-based equity pricing models.[15] Third, Stern Stewart's bold claims stating that EVA explains stock prices considerably better than earnings drew practitioners' and researchers' attention.[16] Despite its recognition among practitioners, EVA involves two major flaws:[17] The first major flaw refers to application of book capital rather than market values for calculating the capital charge

---

[9] See Stern (1994).

[10] Value-based performance measures, marketed by other consultancies such as Marakon Associates, Monitor, Boston Consulting Group, HOLT Value Associates, Alcar, or KPMG, differ by name, but all refer to the RI concept.

[11] See Finegan (1991); EVA® is a trademark of Stern Stewart & Company.

[12] As identified by Michael C. Jensen, see Tully (1998).

[13] E.g., in the first Fortune magazine article on EVA, Tully (1993) describes EVA as "real key to creating wealth" and "today's hottest financial idea and getting hotter".

[14] E.g., Coke stated that "EVA forces you to find ingenious ways to do more with less capital" (Tully, 1993, p. 48). Coke CEO Goizueta further said: "It's the way to keep score. Why everybody doesn't use it is a mystery to me." (Ross, 1996, p. 107).

[15] See Tully (1998); most prominently, Credit Suisse First Boston supported EVA, stating "At CS First, we use EVA as our primary equity valuation tool because it works", (see advertisement of Stern Stewart & Company, in: Fortune, 1997, January 13, p. 10)

[16] "The theory and the evidence all point to the same fundamental conclusion: increasing EVA should be adopted as the paramount objective of any company that professes to be concerned about maximizing its shareholders' wealth." (G. B. Stewart, 1990). "Abandon earnings per share" (Stewart, 1991, p. 2). "The best practical periodic performance measure is economic value added (EVA)" (Stewart, 1991, p. 66). "Earnings, earnings per share, and earnings growth are misleading measures of corporate performance" (G. B. Stewart, 1991, p. 66). "Forget EPS, ROE and ROI, EVA is what drives stock prices" (see advertisement of Stern Stewart & Company, in: Harvard Business Review, 1995, November-December, p. 20).

[17] See Tham (2001c); notably, flaws apply identically to the basic RI measure.

for debt and equity capital and has yet been resolved by a modified EVA measure called 'Refined Economic Value Added' (REVA).[18] The second identified flaw relates to the dependence of EVA on book depreciation schedules: Straight-line or accelerated book depreciation is front-end loaded, ignoring the time value of money.[19]

In particular, for new investments, straight-line book depreciation leads to relatively low EVA in the first years and a growing profile in the following years. For this reason, straight-line book depreciation penalizes the apparent EVA performance of companies with a high growth momentum of the capital employed (and the opposite occurs in companies with a low investment rate and having most of the assets almost fully depreciated - cash cows). Overcoming such distortions from accounting depreciation is central, since biased EVA implies difficulties for management and investors, as follows:

- EVA provides underinvestment incentives to management, referred to as 'old plant / new plant trap', since also significant value-adding investments provide less favorable EVA than retention of old assets.[20]

- EVA involves external communication hurdles to management if, despite the 'old plant /new plant trap', management conducts significant value-adding investments, because deteriorated EVA performance suggests that management destroyed shareholder value.

- EVA is not comparable for firms using different depreciation methods.

- EVA favors equity investments in firms with a relatively old asset base, showing relatively high levels of EVA.

Common value-based performance measures that do not account book depreciation, such as 'Cash Value Added' or 'Economic Margin', generally involve complex computations, adjustments not easy to understand, arbitrary judgments, and/or interpretation difficulties, constituting a major hurdle for external application. [21] Therefore, in 2005, EVA still prevailed as gold standard, measuring value creation for performance monitoring, investor relations, and investment analysis.[22] Nevertheless, bias of EVA due to accounting depreciation remains still a major problem.

---

[18] See Bacidore, Boquist, Milbourn, and Thakor (1997).

[19] See section 3.1.

[20] Empirical studies examining managerial behavior under EVA or similar concepts, e.g., Biddle, Bowen, and Wallace (1999) and Kleiman (1999), provide evidence for implicit divestment incentives.

[21] E.g., CFROI estimates for the Cash Value Added are rather complex, see O'Byrne and Young (2001, p. 407); alternatively, Cash Value Added, following Ottosson and Weissenrieder (1996) and Weissenrieder (1997, p. 5), requires subjective classification of strategic and non-strategic investments; Economic Margin, following Obrycki and Resendes (2000), involves subtraction of a combined charge for cost of capital and depreciation. For a more detailed discussion of depreciation-adjusted measures, see section 3.2.1.

[22] See Stewart (2005).

1.1.2   Inconsistencies and Deficiencies of Prior Empirical Studies

In today's competitive environment, value-relevance of a performance measure, i.e., the empirical evidence that the measure is driving stock prices, serves as important argument to adopt a measure for value-based investor communication and performance monitoring. Regarding this, Stern Stewart heavily promoted EVA as being by far superior to earnings in explaining stock prices.[16] Consequently, researchers extended value-relevance research to include EVA, to examine the accuracy of Stewart's claims.[23]

Testing the value-relevance of performance measures involves relating accounting performance to the market value of the firm and to stock returns.

The former approach derives from financial theory, which suggests that expected future residual income (or EVAs) explain market value.[24] Accordingly, studies conducted by Stern Stewart always relate EVA to market value.[25]

However, most stock valuation models used by equity analysts are mainly focused on stock returns.[26] Additionally, accounting research has widely adopted stock return as criterion for evaluating the information 'usefulness' of performance measures.[27] Starting in the 1960s, studies examined the association of accounting earnings with stock returns, based on the fact that (expected) earnings are of particular interest to investors.[28] Since then, assessment of earnings usefulness became a major objective of accounting research, representing *"the most concerted research effort in accounting history"*.[29] Associations with stock prices are used as operational test of usefulness, since theory suggests that capital markets are efficient and unbiased.[30] Research on earnings showed that associations with returns are persistently weak.[31] In the late 1980s, usefulness studies started to examine cash flows, while RI or RI variants did not yet gain any attention.[32]

Value-relevance research on EVA started with a correlation analysis on EVA and market value, conducted by Stern Stewart in 1991 to substantiate the predominance of

---

[23]   See section 4.1.
[24]   See section 4.2.2.
[25]   See Finegan (1991), O'Byrne (1996), and Stewart (1991); notably, they relate EVA to market value less book value of capital, referred to as 'Market Value Added' (MVA), since theory suggests that market value equals the sum of book value and present value of future RIs.
[26]   See Foster (1986) and Lev (1989).
[27]   See Biddle, Bowen, and Wallace (1997), Cheng, Cheung, and Gopalakrishnan (1993), Easton and Harris (1991), Jacobson (1987), and Lev (1989).
[28]   See Ball and Brown (1968) and Beaver (1968).
[29]   See Lev (1989, p. 153).
[30]   See Ball and Brown (1968, pp. 160-161).
[31]   See Ball and Brown (1968), Beaver (1968), and Lev (1989).
[32]   Research commonly examined usefulness of cash flows (see Bernard & Stober, 1989; R. M. Bowen, Burgstahler, & Daley, 1987; Rayburn, 1986; G. P. Wilson, 1986, 1987).

EVA.[33] However, evidence on the dominance of EVA over competing performance measures has initially been rather vague.[34] Nevertheless, Stewart, co-founder of Stern Stewart & Company, claimed that *"EVA stands well out from the crowd as the single best measure of wealth creation on a contemporaneous basis. [...] EVA is almost 50% better than its closest accounting-based competitor in explaining changes in share-holder wealth"*;[35] and, corporate managers strongly believed in EVA's superior association with stock prices.[36]

In 1996, a regression analysis by Stern Stewart & Co. finally provided support for EVA's dominance over earnings in explaining creation of shareholder wealth:[37] EVA, after-tax net operating profits, and free cash flows explained 56%, 17%, and 0% of excess market value, respectively. However, it may be argued that results were derived from a biased study design as the comparison was made between unadjusted, 'as is' regression models for traditional performance indicators (profit and cash flows) and an EVA model where specifically designed features were enabling better regression results (only the regression model used for EVA allowed a non-zero intercept and included other special features, like: the reference to the respective cost of capital, different regressions for positive and negative values of the performance indicator additional variables and coefficients related to the firm size and to the industry sector).

Therefore, independent academic researchers, not associated with Stern Stewart, reexamined the value-relevance of EVA. Most widely cited in finance and accounting literature, the findings of Biddle et al. refute Stern Stewart's claims on EVA's superiority: Using a consistent regression model across performance measures that includes additional size and industry variables, they find that earnings before extraordinary items are significantly higher associated with firm value than EVA and after-tax net operating profits, explaining 53%, 50% and 49% of variations in market value, respectively; further, EVA does not significantly outperform operating profits.[38] Since Biddle et al. have opened the debate on EVA's value-relevance, numerous

---

33   See Stewart (1991, pp. 215-218), showing that, given groups of 25 firms, changes in EVA divided by invested capital explained 97% of changes in MVA divided by invested capital. Consequently, EVA compensation was claimed to be *"effectively 'self-financing' due to the strength of the correlation between changes in EVA and in shareholder value"* (Chew, Stern, & Stewart, 1995).
34   Finegan (1991) provided a complementary study, correlating EVA ($r = 60\%$), return on capital ($r = 47\%$), growth in cash flow ($r = 20\%$), and earnings-per-share ($r = 10\%$) to Market Value Added. However, correlation results of competing measures are inconsistent, since Finegan compares a measure of total performance, a performance ratio, a performance growth measure, and a per share performance measure to MVA.
35   See Stewart (1994, p. 75).
36   See AT&T's CFO Jim Meenan: *"The correlation between MVA and EVA is very high. So when you drive your business units toward EVA, you're really driving the correlation with market value."* (Walbert, 1994, p. 112); equally, Varity's CEO Victor Rice: *"We fundamentally believe that over time, there is a direct relationship between EVA improvement and a higher share price"* (Rice, 1996).
37   See O'Byrne (1996).
38   See Biddle, Bowen, and Wallace (1997, p. 331); the used regression model accounts for industries, sign of performance, and firm size, and deflates firm performance by beginning capital.

independent studies followed, reconfirming that earnings are indeed superior to EVA in explaining stock prices.[39] However, more recently, studies showed that net earnings no longer predominate other measures, therefore, reopening the debate.[40]

Next to recent inconsistent results, criticism on underlying EVA data further restricts the validity of prior research. Most studies use EVA data provided by Stern Stewart,[41] although *"perhaps the biggest limitation in the preceding studies is the use of publicly available data on EVA values and uses."*[42] Stern Stewart applies few standard adjustments to compute standardized EVA data, rather than custom adjustments to estimate firm-specific EVA, thus, understating the value of EVA.[43] Furthermore, EVA data from Stern Stewart is always based on unadjusted book depreciation, thus, providing always somewhat distorted indications of shareholder value creation.[44] However, also independently computed EVA data applied by other studies does not account for these deficiencies: Used accounting adjustments are quite limited, do not adapt to firm characteristics, and do not include adjustments for book depreciation.[45]

---

[39] See Chen and Clinton (1998), Chen and Dodd (1997, 2001), Fernández (2002a), Kramer and Pushner (1997), and Peterson and Peterson (1996).

[40] Results from Feltham, Isaac, Mbagwu, and Vaidyanathan (2004), Fernández (2002a), Sandoval (2001), and West and Worthington (2004) indicate dominance of EVA, using rather small samples from U.S., Canadian, Spanish, Chilean, and Australian firms. Also, Clinton and Chen (1998) and Tsuji (2006), showing evidence for the outperformance of cash flows, question the dominance of earnings. While Schremper and Pälchen (2001) and Stelter (1999) found evidence for the dominance of CVA.

[41] The majority of studies apply readily available Stern Stewart data to assess EVA, as shown in section 04.1 (see Bacidore et al., 1997; Biddle et al., 1997; Chen & Clinton, 1998; Chen & Dodd, 1997, 2001; Copeland, Dolgoff, & Moel, 2004; G. D. Feltham et al., 2004; Fernández, 2001, 2002a; Finegan, 1991; Kramer & Pushner, 1997; O'Byrne, 1996; West & Worthington, 2004). Notably, any study on U.S. firms, except Peterson and Peterson (1996), applied readily available Stern Stewart data.

[42] See Ittner and Larcker (2001).

[43] *"Studies of EVA's predictive ability typically employ published EVA data estimated by the consulting firm Stern Stewart. [...] This may understate the value of the measures since the published figures exclude the detailed firm-specific adjustments [...]."* (Ittner & Larcker, 2001). Publicly available EVA data provided by Stern Stewart does not include many custom accounting adjustments that Stern Stewart applies for its clients. Adjusted EVA data provided to clients is not published as it represents the core of Stern Stewart's consulting services.

[44] See Stewart (1994, p. 81).

[45] In some instances, studies computed RI and referred to it as EVA, see Ramana (2007) and Singh and Garg (2004) for Indian firms, and Sandoval (2001) for Chilean firms. Otherwise, studies using independently computed EVA data apply a rather small number of adjustments: E.g., Peterson and Peterson (1996), examining U.S. firms, adjust for operating leases, LIFO reserves, goodwill, bad-debt reserves, R&D, and deferred taxes, Heidorn, Siebrecht, and Klein (2001), studying European firms, adjust for operating leases, LIFO reserves, goodwill, R&D, deferred taxes, and extraordinary items, Tsuji (2006), analyzing Japanese firms, adjust for minority interest, provisions, and consolidated adjustment accounts, and Anassis and Kyriazis (2007), examining Greek firms, adjust for R&D, provisions, CIP, and goodwill. However, the EVA concept foresees considerably more potential adjustments, see section 2.3.1.

To conclude, ambiguous results and inappropriate EVA data were suggesting to revisiting the value-relevance of value-based performance measures and represent the starting point from which this study originated.

1.2 Research Objectives

Based on the needs and empirical and theoretical gaps described in the paragraphs above, two research objectives that constitute the scope of this study were defined:

Objective 1: Development of performance measures suitable for value-based management and not affected by bias resulting from book depreciation.

As outlined in section 1.1.1, RI and EVA have the lead over accounting earnings to quantify creation of shareholder value, since RI and EVA both deduct costs for equity capital from earnings, thus, measuring income after charges for all capital employed. It is also important to recognize that, for this reason RI and EVA conceptually dominate accounting earnings from a value-based perspective, independently from any empirical regression evidence. While EVA adjusts for a series of accounting distortions included in accrual-based earnings and RI, this latter, avoiding accounting adjustments, implies less complexity and a simple implementation.

However, RI and EVA imply bias from accounting depreciation. From the researches made in this study, there is no evidence of established value-based performance measures that avoid bias from book depreciation, maintain a tolerable level of complexity, and use publicly available data. Consequently, it has been considered valuable to develop in this study a new performance measure that can be externally applied, describes shareholder wealth creation, and does not suffer from bias introduced by accounting depreciation.

Objective 2: Empirical evidence on associations of prevailing and newly defined performance measures with stock prices.

As described in section 1.1.2, empirical associations of performance measures with market prices are important to justify their relevance for monitoring value-based performance and addressing investor relations, and to understand usefulness in explaining shareholder returns. Therefore, this study examines the value-relevance also of newly developed performance measures. Comparing newly introduced performance measures with their unadjusted equivalents, the empirical study further enables to evaluate Stewart's claim that the depreciation adjustment is not signifi-cant.[46]

Besides, the empirical section of this study includes analysis on the main traditional and value-based performance measures that complements previous EVA studies by

---

[46] To the best of my knowledge, there is no empirical analysis of the depreciation adjustment. E.g., CVA studies do not qualify to evaluate the distortions due to depreciation, since differences between EVA and CVA are not solely due to the depreciation adjustment.

examining independent EVA data from a less biased and sufficiently large sample. In particular:

- EVA data, that in this study has been independently computed from the Thomson financial database and annual SEC (U.S. Securities and Exchange Commission) filings, include a sufficiently large number of accounting adjustments; in this way, derived EVA data is supposed to be more adherent to the actual value creation, thus, increasing its capability to represent actual performances. In contrast, previous research either applies Stern Stewart data including very few unspecified standard accounting adjustments, or estimates EVA data based on few or even omitting accounting adjustments. Possibly, previous dominance of earnings in other studies even resulted from having applied publicly available EVAs.

- The empirical study hereby conducted is based on a sample constituted of a wide number of companies that, while ensuring statistical representation of unspecific general cases, has been depurated of companies and industry sectors whose results are affected by contingency factors and distortions. In particular, the sample includes firms regardless of their trading status or market capitalization, belonging to non-financial sectors, neither heavily affected by technological, medical, and legal issues nor heavily influenced by cyclical factors. Contrarily, commonly applied Stern Stewart data refers to BusinessWeek's league table of 1,000 actively trading U.S. firms, largest in market capitalization. Consequently, prior results were partially affected by specific industry dynamics and bias from overrepresenting large successful companies.

- A relatively high number of 2,147 observations, covering a quite long period of time (1995-2006) that also includes the most recent years, provides a more sound statistical significance to the results of this study with respect to other recent studies, where evidence on EVA's dominance was based on data limited to short periods ending before fiscal year 2000 and mostly based on few firms.[47]

Whilst regressions from O'Byrne (1996) and Biddle et al. (1997) show relatively high R^2s due to the combination of several independent variables, this study avoids using any explanatory variables in addition to the firm's accounting measures. Based on this approach, associations with stock prices are expected to be much 'weaker', but can be exclusively referred back to the accounting performance of the firm. While additional variables, e.g., size and industry variables, may improve 'poor' regression results, they fail to recognize that the overall stock price remains dominated by other non-accounting 'soft' factors driving investors' expectations.

---

[47] Feltham et al. (2004), examining U.S. firms, study 2,608 firm-year observations, however, observations refer to only five fiscal years, namely 1995-1999. Otherwise, Feltham et al. (2004), examining 386 observations from Canadian firms over 1991-1998, Fernández (2002), studying up to 196 observations from Spanish firms over 1992-1998, Sandoval (2001), analyzing up to 372 observations from Chilean firms over 1994-1999, and West and Worthington (2004), examining up to 770 observations from Australian firms over 1992-1998, derive results from rather small samples.

To summarize, there is a central need for a comprehensive, and, above all, unbiased measure of a firm's value addition, given management's and investors' continued scrutiny for shareholder value. This study defines unbiased value-based performance measures, contributing to the further development of value management systems, and tests the value-relevance of new metrics developed in this study and other common performance measures.

1.3   Structure of the Study

The study at hand consists of six chapters, referring to a practice-oriented research problem, as yet described in Chapter 1.

Chapter 2 contains conceptual foundations of the study. In the first part, the concept of value-based management is presented, describing its historic background, providing general definitions and key success factors, and confronting it with stakeholder theory. In the following, a review of value-based performance measures, forming the centerpiece of a value-based management system, includes their definitions, characteristics, uses, and effectiveness. Finally, a detailed description of the EVA method, that still represents the prevailing value-based performance measure, contains a comprehensive calculation guideline and an introduction to its main properties and its application as financial management system, and concludes with a brief discussion.

Chapter 3 starts with a review of the RI and EVA metric, showing that these net residual income measures are biased by accounting depreciation. A discussion of several adjustment techniques demonstrates the need to introduce new unbiased metrics. Thereupon, two depreciation-adjusted performance measures are developed, namely 'Cash Residual Income' (CRI) and 'Cash Economic Value Added' (CEVA), which represent the first major contribution of this study.

In chapter 4, the methodological elements of this study are discussed and derived. First, a discussion of previous empirical studies on value-based performance measures shows some ambiguity in the results and a notable research gap. Accordingly, a study design for the following empirical study is derived, including four underlying research hypotheses, two model specifications regarding associations with market value and returns, relative and incremental information content tests, the underlying sample of 2,147 firm-year observations, and various independent and dependent variables.

Chapter 5 commences with a case study that shows the computation of the newly developed value measure described in this study and the ones currently mandated, in order to illustrate new measurement concepts, to provide a more detailed outline of independently calculated custom EVA data, and to demonstrate the interrelationship among various variables of the subsequent study. After that, follows an empirical examination of derived research hypotheses, constituting the second major contribution. At first, summary and correlation statistics on pooled panel data serve to verify underlying data and to validate presumptive relationships. In the following, a regression analysis examines associations between measures of the firm's operating performance with stock prices, based on two models: In the first model accounting performances have been associated with market values of the firms, as derived primarily from the discounted cash flow valuation model. In the second model, absolute and incremental deflated values of accounting performance are related to stock returns beyond market level returns. Based on these two models, all variables are tested for relative and incremental information content. Additionally, a sensitivity analysis examines the robustness of the basic results.

Chapter 6 concludes this study elaborating on: first, implications for corporate managers, accounting policymakers, and the investment community, second, limitations concerning the contents and applied methodology of this study and, finally, possible areas for further future researches.

The following figure outlines the structure, as described above.

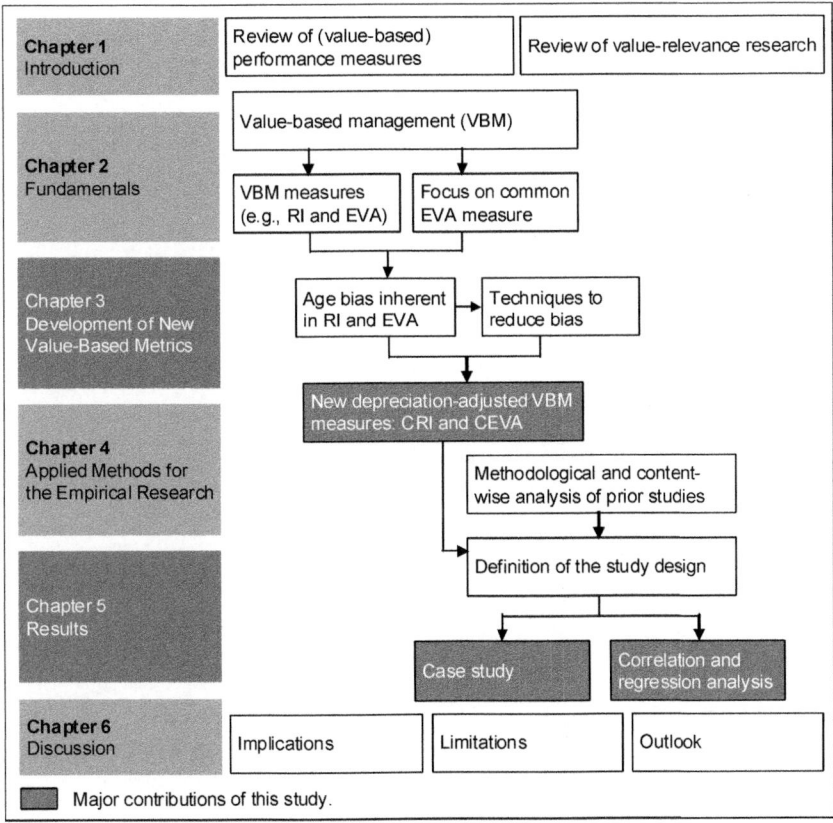

Figure 1   Structure of the Study

2 Fundamentals

In this chapter, an introduction to the concept of 'Value-Based Management' and value-based performance measurement provides the definitional basis of the following study.

Thereto, the origin, scope and realization of value-based management are presented. Correspondingly, value-based performance measurement methods (namely, the 'metrics'), representing the core of a value-based management system, are defined and evaluated. Then, an overview follows about different uses of value-based performance metrics and their effectiveness if applied to determine management compensation. To deepen the conception of value-based measures of corporate performance, a more detailed introduction to the 'Economic Value Added' measure joins, highlighting its computation and applications.

This study reviews value-based performance metrics that can be adopted within a value-based management system and externally applied to monitor the firm's additions to shareholder wealth. Therefore, it excludes expectations-based metrics, since these do not fulfill with accountability principles as their results can be arbitrarily manipulated and are driven by market expectations. [48] Moreover, it disregards market measures, which are heavily influenced by exogenous factors uncontrollable by the management.[49] Finally, the application of performance measures for firm valuation is limited to an empirical study of value regressions, whilst models aimed at predicting market values are beyond the scope of this study. [50]

---

[48] E.g., 'Expected Improvement' (O'Byrne & Young, 2006) or 'Expectations Based Management' (Copeland et al., 2004) both define firm performance as actual performance less market expectations, to explain a higher share of shareholder returns

[49] Frequently, consultancies introduce specific market measures as pendant to the internal VBM measure: E.g., 'Market Value Added' (MVA) corresponds to the internal EVA measure, or Total Shareholder Return (TSR) corresponds to the internal TBR measure.

[50] See, e.g., Madden (1999, pp. 161-187) on the integration of growth and fade to evaluate firms based on 'CFROI', and Abate and Grant (2001, pp. 45-65) on diverse models to evaluate firms based on 'Economic Profit'.

2.1   Value-Based Management (VBM)

2.1.1   Emergence of VBM Responding to Increasing Shareholders' Pressure

Until the beginning of the 20th century, firmowners were usually also leading key management positions. With the development of the stock exchange and of large corporations, firm's shareholders were more and more differing from the firm's management, with the possibility for the latter to take decisions that benefit themselves and other stakeholders at the expense of the firm's investors.[51] Adequate corporate governance structures are nowadays generally adopted in order to mitigate such risks. Nevertheless, these became common only in the last decades, as before firms traditionally considered shareholders as one of various stakeholder groups without prominently pursuing their interests. In such context, institutional shareholders passively bought or sold shares without being actually involved in assessing managerial performance and issues of portfolio companies.

However, unequaled takeovers and shareholder activism appeared over the last decades. This pressure from capital markets and shareholders initiated a fundamental change in corporate management to pay more attention to shareholder needs, giving rise to 'Value-Based Management' (VBM). The following more detailed description of these market dynamics allows a better comprehension for the conditions under which this strategic management approach emerged.

In the 1980s, a large number of sizeable leveraged buyouts emerged as first source of managerial discipline. Buyout companies acquired underperforming firms, sold unprofitable divisions, reduced the workforce, and re-listed firms in a leaner, more competitive form with a new management more focused in creating value to shareholders. Additionally, shareholders became increasingly active in protecting their rights; exemplarily, as response to increasing greenmail and anti-takeover activities, in 1985, shareholders formed the Council of Institutional Investors to defend their interests.

While the junk bond market provided tremendous funding for buyouts, its collapse caused a substantial decline in acquisitions of publicly traded corporations, falling from 462 in 1988 to 148 in 1991.[52] In the 1990s, an unprecedented level of shareholder activism made up for the decline in takeovers and constituted the second source of managerial discipline. Primarily institutional investors actively targeted under-performing portfolio firms to improve their performance by replacing and addressing management towards a more focused value-oriented organization and strategies. This behavioral change is attributable to three sources:

---

[51]   This conflict of interest is referred to as agency problem, arising in corporations with public float due to the separation of ownership and control and asking for a series of corporate governance mechanisms to protect shaerholders, see Berle and Means ([1932] 1968).

[52]   See Mergerstat (1993).

- First, institutional shareholders gained substantial ownership, largely due to funding of pension obligations based on the Employee Retirement Income Security Act, passed in 1974. Thus, the growth in institutional holdings of U.S. firms' common stock from USD 673 billions in 1970 to USD 15 trillions in 1998, with pension funds' holdings increasing from USD 213 billions to 7 trillions, increased the power of shareholders.[53]

- Second, pension funds commenced applying an indexing strategy for management of a significant share of assets, limiting passive trading opportunities of fund managers. E.g., in 1998, public pension funds indexed on average 63% of assets.[54]

- Third, revised proxy solicitation rules, following passage of SEC (U.S. Securities and Exchange Commission) Rule 14a-2(b) in 1992, resulted in substantial alleviation of communication among shareholders. As a result, for example, the Council of Institutional Investors started distributing a focus list on poorly performing firms among members.[55]

Next to the emergence of a takeover market and active shareholders, several other developments required companies to be competitive not only in commercial markets but also in capital markets:[56]

- Globalization and deregulation of capital markets

- Higher liquidity of securities market

- Advances in information technology

- Improved capital market regulations

- Increasing investor attitude towards higher remuneration, risk management and longer investment horizons

- Increasing role of institutional investment funds

- Recognition of a missing linkage of traditional accounting measures to equity value

- Reports on shareholder returns in the business press

Consequently, shareholders and the financial community (with particular reference to professional fund managers, analysts and directors), determined increasing pressure on managers and value-based compensation incentives[57], whilst questioning expansion and divestment of business activities, and strongly criticizing doubtful disclosure practices, excessive management compensation, firms' reluctance to address analysts' questions and concerns, and boards of directors with biased composition.

---

[53] See The Conference Board (1999).
[54] See The Conference Board (2000).
[55] The Council of Institutional Investors provides the focus list online to general members, see www.cii.org.
[56] See O'Byrne and Young (2001, pp. 5-8) and Rappaport (1986, pp. 2-6).
[57] See Martin and Petty (2000, p. 22).

To address shareholders' needs, corporations commenced to prioritize value creation for shareholders, as suggested by VBM. Therefore, VBM has yet been commonly accepted in practice, when it was prominently introduced by VBM pioneer Rappaport in 1986.[58]

2.1.2 Maximization of Shareholder Value as an Overall Objective

Under VBM, the overall mission or strategic goal of a firm is to maximize shareholder value.[59] The firm's creation of shareholder value equals total stock returns, i.e., dividends and price appreciation of the stock, adjusted for changes in capital structure and share issuances.[60]

To manage shareholder value creation, the VBM concept uses value drivers that link the strategic corporate objective of shareholder value maximization to any management decision.[61] Rappaport illustrated the derivation and use of value drivers in a four-step VBM framework titled 'Shareholder Value Network', see Figure 2.[62]

---

[58] "The principle that the fundamental objective of the business corporation is to increase the value of its shareholders' investment is widely accepted." (Rappaport, 1986, p. 1).

[59] "Shareholder value and VBM have become catchwords to describe this philosophy of value maximization." (Wallace, 2003, p. 121).

[60] See Rappaport (1986, p. 1). Definition of shareholder value creation based on stock returns deliberately uses market prices, which reflect the consensus of all stockowners, although individual perceptions of value vary due to differing quality of information, perception of control, time horizon, uncertainty and tolerance for risk, see Knight (1998, pp. 21-29). Notably, definitions on shareholder value often focus on stock quotations, not available for non-public firms (see Knight, 1998, p. 21; Rappaport, 1986, p. 11). However, VBM can be equally adopted by private firms (Finegan, 1991).

[61] "Properly executed, VBM is an approach to management whereby the company's overall aspirations, analytical techniques, and management processes are all aligned to help the company maximize its value by focusing management decision making on the key drivers of value." (Copeland, Koller, & Murrin, 1994, p. 93)

[62] Rappaport (1986, pp. 50-77) provides a generic framework for decision making in line with VBM. Martin and Petty (2000, pp. 66-70) discuss the same methodological approach, providing a corporate example linking value drivers to business unit value.

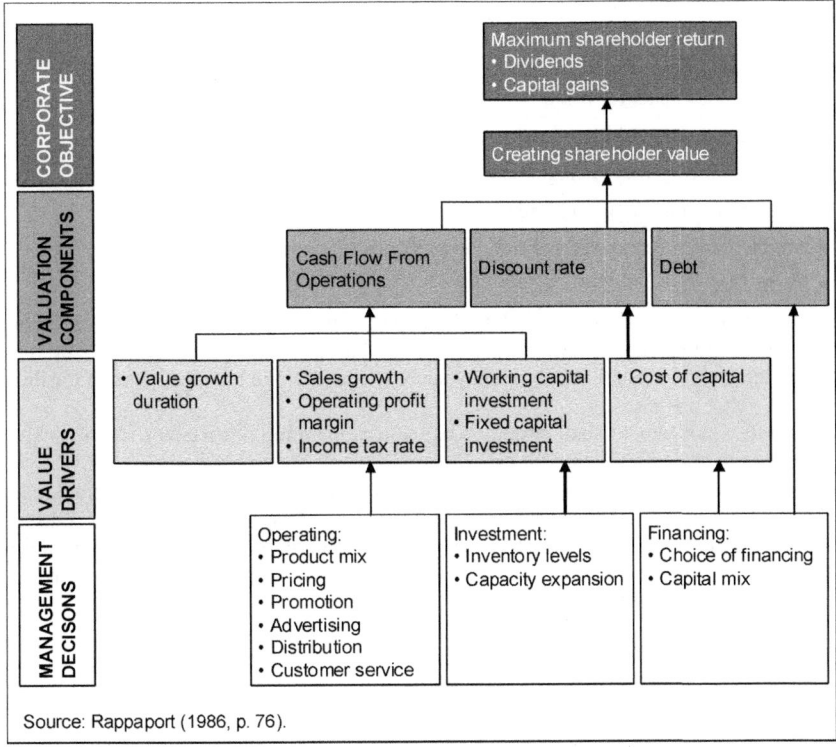

Figure 2   Shareholder Value Network

First and foremost, the firm aims to maximize shareholder wealth, providing attractive long-term stock returns. [63] Correspondingly, the firm's value creation is measured internally, e.g. adopting a free cash flow valuation.[64] Second, quantification of shareholder value is disaggregated into valuation components, in this case: Cash flows, the discount rate, and the amount of capital. Third, value drivers, i.e. critical factors that

---

63 Alternatively, shareholder value added can be defined via increases of market-to-book ratios, see Day and Fahey (1990, p. 158).
64 Rappaport (1986, p. 12) suggests the cash flow analysis; the cash flow perspective is often adapted to assess individual projects or business units. Free cash flow valuation adds the present value of free cash flows to the value of non-operating assets such as marketable securities, excess real estate, or over-funded pension plans, and subtracts future claims, such as interest-bearing debt, capital lease obligations, under-funded pension plans, and contingent liabilities. Nevertheless, residual income measures can be equivalently applied. For an overview of applicable value-based performance measurement concepts, see section 2.2.2.

affect valuation components, directly connect decision-making and firm value and facilitate to understand, define, and communicate value-adding activities.
Value drivers ideally fulfill the following criteria:

- Significant impact on value

- (Monthly) measurable

- Controllable by line management

- Aligned with decision-making processes

- Supplemented with targets and responsibilities

- Periodically reviewed

Following the cash flow model, Rappaport suggests adopting growth duration, growth, margins, tax rates, investments and the cost of capital as value drivers. Complementary, scenario analysis may serve to identify and manage interrelationships among value drivers.[65]

Fourth, management shall systematically align operating, investment, and financing decisions to maximize value drivers.

Contrary to a top-down management process, VBM involves integrating employees from all hierarchical levels and organizational functions, since shareholder value is created or destroyed throughout the corporation. Therefore, VBM adoption requires focusing on value creation at all levels, prominently demonstrated as a pyramid, see Figure 3.[66]

---

[65] See Copeland et al. (1994, pp. 103-109).

[66] *"VBM lays the foundations for the development of mission and subsequent corporate and individual plans and goals by enabling managers to address and resolve unavoidable dilemmas, [...] toward building a high growth organization in which individual performance improves and heightened individual achievement drives economic success."* (C. Anderson, 1997, p. 25). *"It focuses on better decision making at all levels in an organization.* (Copeland et al., 1994, p. 93).

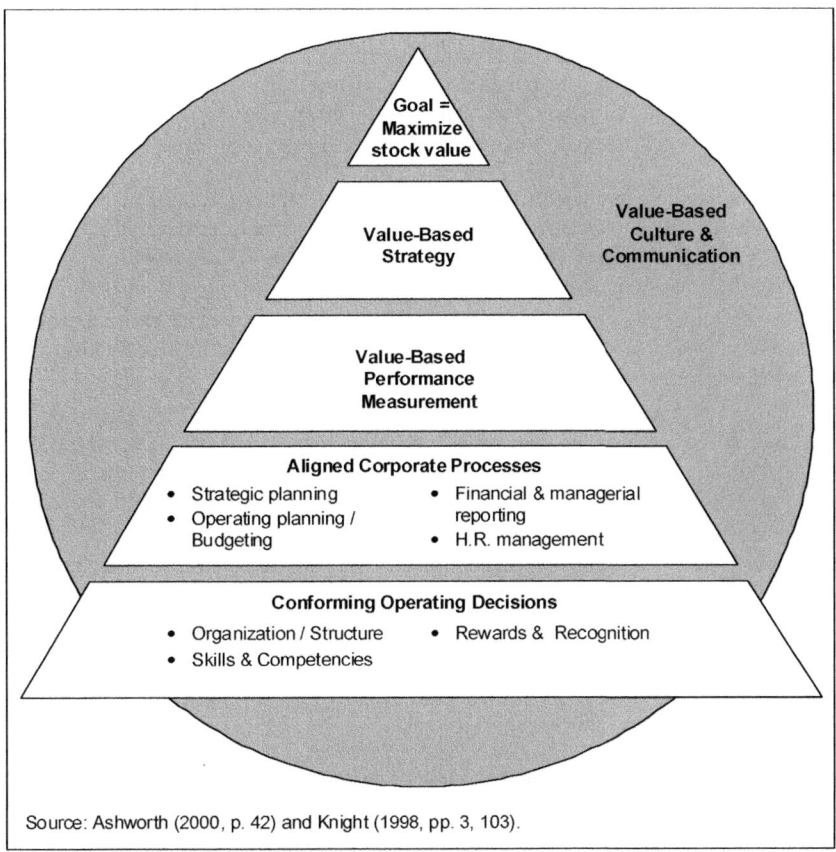

Source: Ashworth (2000, p. 42) and Knight (1998, pp. 3, 103).

Figure 3   Value-Based Management (VBM) Pyramid

At the top of the pyramid, shareholder value creation represents the fundamental purpose or mission of the firm. Then, value-based performance measurement underpins the strategy and provides guidance for all corporate processes. Resulting from aligned corporate processes, operating decisions enable shareholder value maximization, building the fundament of the pyramid. Finally, ensuring that all strategic and operating decisions are aligned with the shareholders' interests, VBM transforms the corporate culture.[67] Consequently, VBM facilitates communication and

---

[67] *"VBM can best be understood as a marriage between a value creation mindset and the management processes and systems that are necessary to translate that mindset into action."* (Copeland et al., 1994, p. 97). *"VBM instills a mind-set where everyone in the organization learns*

cooperation among various divisions and departments, linking strategic planning with operating divisions and finance and providing a common language for employees.

As yet illustrated by the pyramid, VBM requires adopting a shareholder value perspective in all corporate functions, since these are mutually dependent. In particular, four management processes govern the cycle of value creation, as shown in Figure 4:

- First, corporate board and business units develop a strategy to maximize firm value and translate it into short-term and long-term performance targets, which shall be based on value drivers, tailored to the firm, and ideally expressed in terms of value.[68]

- Second, budgeting defines resource allocations to achieve targets of the succeeding calendar year, weighing and balancing the value added for shareholders from competing investment scenarios.[69]

- Third, value-based performance measurement monitors performance against targets. Value-based performance measures are often considered the core of a VBM system.[70] Internally communicated, performance measures reinforce the execution of strategic plans.[71] Externally applied, performance measures support an investor relations communication strategy.[72]

- Fourth, incentive systems, defining total and variable compensation based on reported performance, encourage employees to meet targets.[73]

---

to prioritize decisions based on their understanding of how those decisions contribute to corporate value [...] creating management alignment by linking [...] strategic planning, financial reporting, and compensation / incentive planning." (Knight, 1998, p. xiii).

[68] See Copeland et al. (1994, pp. 110-112) and Knight (1998, pp. 112-114).

[69] See Copeland et al. (1994, p. 113) and Knight (1998, p. 115).

[70] *"VBM calls on managers to use value-based performance metrics for making better decisions. It entails managing the balance sheet as well as the income statement, and balancing long- and short-term perspectives."* (Koller, 1994, p. 87). *"VBM is a strategic performance measurement initiative organizations embark on to focus internal performance and incentives on value creation. [...] It is dependent on a firm's financial accounting and reporting as a starting point in computing the metrics used. [...] However, the business strategy must be designed to maximize value creation."* (Frigo, 2002, p. 6).

[71] *"The old truism states, 'What gets measured gets done'. [...] Value measures can be used in reporting to reinforce the managing-for-value message."* (Knight, 1998, p. 118).

[72] *"VBM involves [...] managing the investor community, [...] developing optimum value creating strategies, [...] and delivering shareholder value through integrated performance management [...] to create and deliver shareholder value."* (Ashworth, 2000, p. 44).

[73] *"It is easy to dismiss performance measurement as something that accountants do, but we are always amazed at how carefully managers scrutinize the numbers on which they are evaluated."* (Copeland et al., 1994, p. 113)

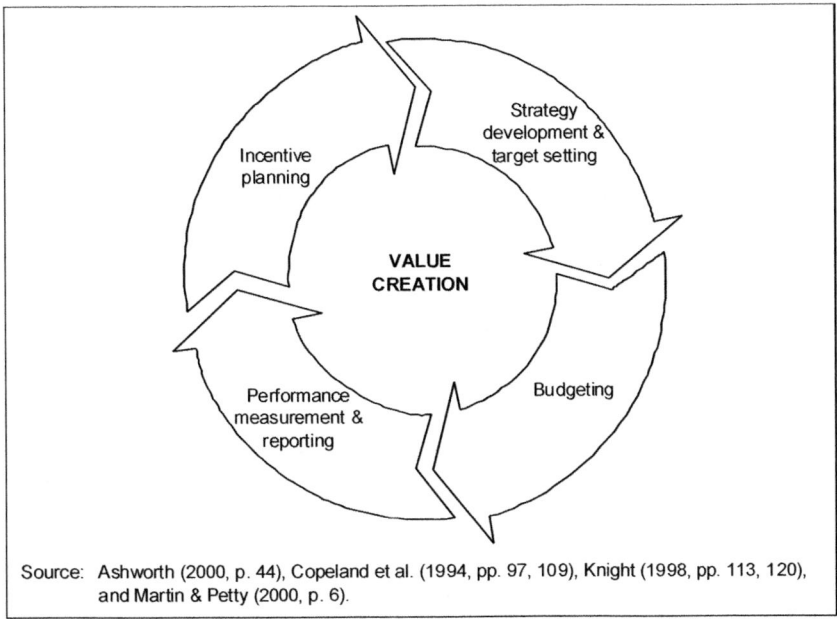

Source: Ashworth (2000, p. 44), Copeland et al. (1994, pp. 97, 109), Knight (1998, pp. 113, 120), and Martin & Petty (2000, p. 6).

Figure 4   Cycle of Value Creation for Shareholders

Notably, VBM is commonly understood as the linkage of management compensation to value-based performance metrics.[74]

Before considering VBM, firms commonly face an internal conflict between strategic planning, which emphasizes on long-term growth, and annual budgets, which emphasize on short-term earnings per share. Additionally, firms may be exposed to threats from increasingly competitive capital markets (see section 2.1.1), suffer from inefficient and inadequate structures, and lack instruments to monitor and communicate its value addition. In this situation, VBM, once properly implemented, provides primarily four internal benefits to corporations:[75]

---

[74] *"The main role of VBM in the corporate management process is to develop performance measures linked to shareholder value and, in some cases, to design incentive compensation plans based on such measures."* (Wallace, 2003, p. 122). *"The fundamental premise upon which VBM systems are based is that to sustain the wealth creation process, managerial performance must be measured and rewarded using metrics that can be linked directly to the creation of shareholder value. Thus, the marriage of value based performance metrics and incentive compensation is the very heart of the VBM programs [...]."*(Martin & Petty, 2000, p. 6).

[75] See Knight (1998, pp. 45-46).

- First, VBM balances short- and long-term trade-offs, sets management priorities, and improves resource allocation by incorporating the idea of shareholder value maximization into the planning process.[76]

- Second, VBM, by focusing the management on long-term equity value and providing valuable tools for assessing the impact of company strategies on corporate value creation, ideally avoids stock undervaluation and hostile takeovers, increases the value of stock options, and facilitates the use of stock for mergers and acquisitions.

- Third, VBM represents a recognized mean to enable change by reviewing and realigning strategic and operating plans and all corporate processes.

- Fourth, VBM assists in communicating the firm's value addition within the firm and to the investment community by installing a value-based performance measurement system that summarizes its creation of shareholder value.

Consequently, the VBM method is considered one of the most significant contributions to corporate financial planning in the last two decades, integrating financial models with strategic economic philosophy.[77]

2.1.3 Critical Success Factors for VBM Implementation

Implementation of a VBM system is time-consuming and, therefore, costly: The design of a VBM system takes at least six months.[78] The whole implementation takes about two years.[79]

In this respect, a brief review of implementation practices reveals five critical success factors of VBM adoption.
- Management commitment
- Integrated corporate processes
- Autonomous and accountable business units
- Extensive training
- Aligned compensation

Management commitment

One common characteristic of successful VBM firms is that the firm's management understands the VBM concept and its implications in detail and, most of all, shows passionate support of VBM.[80]

---

[76] See Rappaport (1986, p. 1).
[77] See Mc Taggart, Kontes, and Mankins (1994, pp. 4-6).
[78] See Ross (1997, p. 116).
[79] See Copeland et al. (1994, p. 117).
[80] See O'Byrne and Young (2001, p. 93).

There is also empirical evidence that the extent of management commitment drives the successfulness of VBM adoption:[81] A survey finds that 67% of firms with an explicit commitment to shareholder value maximization show a high impact of VBM adoption on the relative share price, while just 30% of firms with an implicit commitment to value exhibit a high price increase.

Presumably, management's enthusiastic and explicit commitment to VBM and, therefore, affects the mind-set of employees, builds confidence in management, and introduces cultural change. Accordingly, stock performance reflects to some extent changes in corporate culture, demanded by capital markets.

Integrated corporate processes

Disintegration of corporate processes seems to be a major driver to implement a VBM system based on one comprehensive value-based performance measure. [82] Additionally, empirical evidence supports that in particular the design of the performance measurement system and budgeting process affects the successfulness of VBM adoption:

A survey found that value-based performance measurement systems seldom span all hierarchical levels:[83] Performance measures are primarily used at the corporate level, to a lesser extent at major business units, and to an even lesser extent at functional areas within major business units. Consequently, VBM methods are often not pushed down in firms, presumably due to higher costs of implementation or complex or subjective definitions of value-based performance measures. However, limitation of value-based performance measurement to accounting and control systems is often associated with unsuccessful VBM implementation. [84] On the contrary, successful VBM adopters primarily invest in corporate-wide performance measurement systems, which also incorporate value drivers, and, additionally, integrate budgeting with strategic planning. Accordingly, an empirical study found that successful VBM firms have implemented an effective integrated budgeting process within their strategic-

---

[81] See Haspeslagh, Noda, and Boulos (2001, p. 4). The authors evaluate successfulness of VBM implementation by dividing 117 VBM users into three groups based on self-reported share price improvements, which are validated by comparing the three-year average annual total shareholder return before and after VBM implementation. The validation showed statistically significant results confirming the self-reported statements on share price improvements.

[82] See a comment by SPX's financial planning and analysis director Charles Bowman: *"We had a financial management and planning system that was unclear, complicated, disjointed – we needed to reform it. We had an incentive compensation system that had been frequently modified, which was hurting its credibility – our people referred to it as the 'flavor of the month'. And we had some internal frustration between financial objectives versus accounting objectives. So that's really what led up to it."* (Ross, 1997, p. 116).

[83] See Ryan and Trahan (1999, p. 52); generally, the survey of 162 chief financial officers from large U.S. industrial and non-financial services firms provides evidence based on 328 observations from 112 DCF users, 76 EVA users, 59 ROIC users, 56 CFROI users, and 25 users of other value-based performance measures.

[84] See Haspeslagh, Noda, and Boulos (2001, p. 7).

planning systems[85] in a proportion more than double with respect to unsuccessful VBM firms Consequently, integrated budgeting no longer represents a short-term, day-to-day management instrument independent from long-term strategic plans, but supports maximization of shareholder value.

Autonomous and accountable business units

The autonomy of business units is another characteristic that successful VBM adopters frequently share; i.e., business units are neither organized as a matrix organization nor share common resources.[86]

In this sense, successful VBM adopters commonly provide upfront funding to business units, promoting sovereign and decentralized decision-making: A study found that 52% of successful and just 17% of unsuccessful VBM firms have corporate centers funding complete strategies rather than discrete operating projects. [87] Consequently, upfront funded business units are empowered rather than controlled by the corporate, which enables business unit managers to focus on implementing rather than justifying value-creating activities. Following, empirical evidence suggests that low involvement of the corporate center in operation decisions adds to the successfulness of VBM implementation.

Additionally, experience shows that business heads of successful VBM adopters are usually held accountable over the long term due to extended job tenures.[88]

Extensive training

A study shows that training on applied value-based (measurement) methods is commonly provided:[89] 84% of surveyed VBM firms conduct trainings during VBM implementation, of which 84% conduct training internally rather than hiring an outside specialist.

However, it is especially critical to ensure that all employees are familiar with used methods. Experience shows that it is substantially driving success to train all corporate managers and all key employees:[90] 62% of successful VBM firms and only 27% of unsuccessful VBM firms trained more than 75% of managers on VBM methods. Furthermore, 45% of successful VBM firms and just 13% of unsuccessful VBM firms trained more than 25% of all employees in VBM methods.

---

[85] See Haspeslagh, Noda, and Boulos (2001, p. 7).
[86] See O'Byrne and Young (2001, p. 92).
[87] See Haspeslagh, Noda, and Boulos (2001, p. 9).
[88] See O'Byrne and Young (2001, p. 93).
[89] See Ryan and Trahan (1999, p. 53); whereby, observations are reduced for analyses on employee training due to some incomplete responses.
[90] See Haspeslagh, Noda, and Boulos (2001, p. 5).

Aligned compensation

Compensation plans are commonly used to increase employees' ownership in the firm. Successful VBM firms typically tie management compensation strongly to value creation in business units, while avoiding emphasis on stock options.[91]

In addition, empirical evidence highlights that value-based compensation plans shall preferably cover a large share of employees:[92] While 53% of successful VBM firms include more than 50% of employees in bonus programs, only 24% of unsuccessful VBM firms provide value-based incentives to more than half of their employees.

Overall, it seems that the design and coverage of compensation plans rather than absolute compensation levels contribute to the successfulness of VBM implementation.

2.1.4   Validity of VBM Systems within Sustainability and Stakeholder Models

In the first instance, VBM seems to give preference to shareholders at the expense of other interest groups, leaving socially responsible activities to governments. Alternatively, 'Stakeholder Theory' and 'Corporate Sustainability' integrate a multiplicity of interests rather than privileging shareholders. However, the following discussion shows that these alternative approaches do not contest the validity of the shareholder approach.

Contrary to the shareholder value concept that focuses exclusively on shareholders' needs, 'Stakeholder Theory' suggests considering the interests of all stakeholders, such as shareholders, employees, customers, suppliers, competitors, the financial community, environmental interest groups, the local community, and the federal government. Such concept is based on the assumption that a fair treatment of all stakeholder groups results in more highly motivated employees and a superior supply chain, increasing sales and margins, and, therefore, improves the firm's value creation to shareholders.[93] Therefore, the stakeholder approach evaluates the success of a firm based on its contributions to all its stakeholders. Contrarily, the shareholder concept suggests assessing firm performance only based on its value additions to shareholders.

Stakeholder theory yet extends the firm's responsibility from shareholders to all its primary stakeholder groups. The 'Corporate Sustainability' concept even goes further, widening corporate responsibility to include the management of impacts upon the environment and society, see Figure 5.

---

[91] See O'Byrne and Young (2001, pp. 92-93).
[92] See Haspeslagh, Noda, and Boulos (2001, p. 6).
[93] See Freeman (1984) and Donaldson and Preston (1995) for a general introduction and discussion of the stakeholder theory; for a classification and assessment of shareholders, see Mitchell, Agle, and Wood (1997).

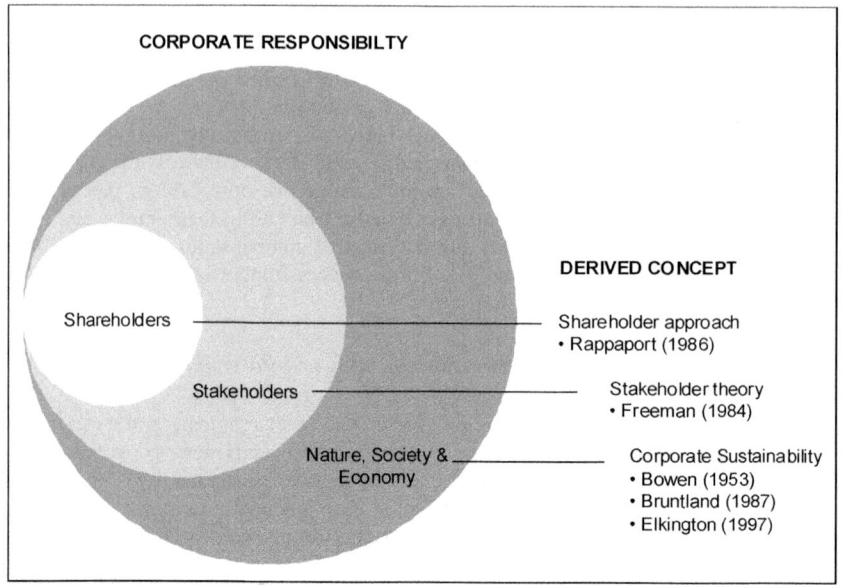

Figure 5   Definitions of Corporate Responsibility

The sustainability approach refers to the concept of 'Sustainable Development', which introduced the idea that firms have a common societal goal, namely to focus co-equally on environmental, social, and economic performance.[94] In addition to the demand for sustainable development, the sustainability concept integrates arguments from three other concepts:[95] The 'Corporate Social Responsibility' concept contributes ethical arguments reasoning why firms have an obligation to assist society from a moral philosophical view.[96] Next, 'Corporate Accountability Theory' provides ethical arguments that firms shall issue reports on sustainability.[97] Finally, stakeholder theory adds business arguments as to why firms should aspire to sustainable development from a strategic management perspective.

More recently, corporate sustainability has rapidly gained ground among businesses and investors: The percentage of Global Fortune Top 250 firms issuing environmental,

---

[94] „A process of change in which the exploitation of resources, the direction of investments, the orientation of technological development, and institutional change are all in harmony and enhance both current and future potential to meet human needs and aspirations."(Bruntland, 1987).

[95] See M. Wilson (2003).

[96] See H. R. Bowen (1953).

[97] See Elkington (1997) introducing 'Triple Bottom Line' reporting on environmental, social, and economic performance.

social or sustainability reports in addition to their financial reports increased from 35% in 1999 to 52% in 2005.[98] In 2001, the European Commission published recommendations for the integration of sustainability issues in corporate annual reports. [99] Furthermore, Dow Jones introduced global, North American, and European sustainability indexes in 1999, 2005, and 2006, responding to investors that increasingly perceive sustainability as essential success factor and, thus, increasingly investing in firms with best sustainability practices.[100] This increasing popularity of the sustainability model, which yet integrates stakeholder theory, seriously questions the validity of VBM.

Furthermore, stakeholder theory directly confronts VBM, reasoning that an exclusive focus on shareholder value may come at the expense of other stakeholders, e.g. employees being laid off due to downsizing, and may damage the firm's reputation, thus, ultimately hurting financial performance and reducing shareholder value.[101] Likewise, the sustainability model challenges VBM based on the same logic. However, there is no clear empirical evidence on the relationship between the firm's social responsibility, reputation, and market value: The relationship between a firm's social responsibility and financial performance has not been empirically supported. [102] Although there is evidence on the positive relationship between shareholder value creation and reputation, empirical results do not allow inferring causality between value creation and stakeholder benefits.[103]

Conversely, VBM theory argues that maximization of shareholder value leads to overall higher firm value and, thus, more resources available to allocate to all stakeholders. Therefore, VBM firms, which put capital to its best use and maximize long-term returns, benefit society.[104] Thus, VBM automatically satisfies the interests of stakeholders other than shareholders, environmental associations, and governmental bodies, although the interests of those groups are neither specified nor especially pursued. Indeed, prior to the era of VBM, yet Adam Smith's notion of an invisible hand suggested that the maximum public benefit automatically results from investors seeking their maximum private gain.[105]

---

[98] See Kolk, van der Veen, Pinkse, and Fontanier (2005).
[99] See European Commission (2001). Commission Recommendation of 30 May 2001 on the recognition, measurement and disclosure of environmental issues in the annual accounts and annucal reports of companies (2001/435/EC). *Official Journal of the European Communities* L 156 (13 June 2001): 33–42.
[100] See ‚Dow Jones Sustainability World Indexes Guide' (version 9.1, January 2008, p. 5).
[101] See Black, Carnes, and Richardson (2000) for empirical evidence, showing a significant positive association between a firm's reputation and its market-to-book ratio.
[102] See Waddock and Grave (2000).
[103] See O'Byrne and Young (2001, p. 14).
[104] See von Hayek (1969).
[105] *"Every individual endeavors to employ his capital so that its produce may be of greatest value. He generally neither intends to promote the public interest, nor knows how much he is promoting it. He intends only his own security, only his own gain. And he is in this led by an invisible hand to promote an end, which has no part of his intention. By pursuing his own interest he frequently promotes that of society more effectually than when he really intends to promote it."* (Smith,

Apart from that, VBM, stakeholder theory and corporate sustainability complement each other, as demonstrated by empirical evidence: On the one hand, firms with higher value creation rather than social responsibility have stronger non-financial reputations; on the other hand, increasing stakeholder benefits results, to a certain degree, in additional shareholder value.[106]

Compatibility is further underlined by conceptual differences, as follows: On the one hand, the stakeholder and sustainability concept describe an overall management strategy, but lack a rule for trade-offs among competing interests.[107] On the other hand, VBM provides directions on how to allocate resources among competing interests, but it does not provide management with a specific strategy for maximizing shareholder value. In this respect, the sustainability concept may aid VBM firms in enhancing shareholder value.[108]

To conclude, financial success and shareholder value creation in the sense of VBM are compatible with stakeholders', environmental and societal interests, if not prerequisite to enhance multiple-stakeholder, ecological, and social benefits.

2.2 Value-Based Performance Measures

Performance metrics are far more than just a controlling instrument, as stated by Jensen and Meckling (1999, p. 8): *"Performance measurement is one of the critical factors how individuals in an organization behave."* A performance metric, which quantifies the firm's value creation, represents the centerpiece of a VBM system; consistently adopted across organizational functions and hierarchical levels, value-based performance measures improve financial information to management and investors, support strategic and operating decision-making, and align managerial behavior with shareholder objectives.[109]

However, traditional performance measures imply several shortcomings for value management, shown as follows.

1776). However, Anderson (1997, pp. 30-31) notes that Smith's concept of the invisible hand involves three fundamental flaws: It does neither incorporate the competing interests of competitors, nor self-interest due to increased competition nor an underlying moral theory.

[106] See Wallace (2003).

[107] While, e.g., mapping the strength of stakeholders' interests and stakeholders' power provides some decision input, it does not solve conflicts of interests among stakeholder groups. Consequently, Henderson (2001, pp. 21-22) states that *"there is no solid and well-developed consensus which provides a basis for action"*.

[108] *"Corporate Sustainability is a business approach to create long-term shareholder value by embracing opportunities and managing risks deriving from economic, environmental, and social developments."*(,Dow Jones Sustainability World Indexes Guide', version 9.1, January 2008, p. 7).

[109] See Stewart (1991).

2.2.1   Unsuitability of Traditional Performance Measures for VBM Scopes

In prior centuries, output, volume, products sold, sales growth and income growth described firm performance. Metrics that are derived from accounting systems did not gain attention from management and investors, although double-entry bookkeeping has yet been devised in 1494.[110] Then, accounting measures gained recognition after World War II in order to manage complex multiple product businesses. Contrary to non-financial performance measures, accounting measures are readily available and comparable among differing businesses.

While traditional earnings and return measures are commonly used in the financial community, they are inadequate from a value-based view, not reflecting creation of shareholder wealth appropriately.[111]

Above all, cash flows and not earnings determine the market value of the firm and, thus, additions to shareholder wealth, being generally the capital goodwill (increase of stock prices) a substantial part of the shareholder return that is considered by investors in addition to dividend expectations. Specifically, the amount, timing, and risk of expected future cash flows determine firm value.[112] Single-year reported earnings do not measure value creation.[113] Furthermore, management commonly manipulates earnings via the timing of revenue recognition, inventory re-evaluations, deferred charges, and further accounting methods, to fulfill earnings targets expected by investors.[114] However, 30 years of stock market data show that the price of a stock with manipulated earnings tends to fall to the level derived from cash flows in the long term.[115] Consequently, not only valuation theory but also management practices and market dynamics disqualify earnings as objective indicator of the firm's shareholder value creation.

Specifically, the following shortcomings describe why rewarding earnings or earnings growth may lead to value-reducing decisions inconsistent with shareholder interests:

- Earnings include non-cash items such as depreciation & amortization with no impact on shareholder value, even though such items some times are ideally considered as a reference level for replacement capital expenditure (CAPEX) needs.

- Earnings contain distortions from changes in accounting principles, not indicating creation of shareholder wealth.

---

[110] See Pacioli (1494)
[111] Various authors outline the inadequacy of traditional performance measures (e.g. Grant, 2003, pp. 50-52, 145-167; Martin & Petty, 2000, pp. 35-46; Rappaport, 1986, pp. 20-27; G. B. Stewart, 1991).
[112] See section 2.2.2.
[113] See section 4.2.2.
[114] See Knight (1998, pp. 174-175).
[115] See Millman (1997).

- Earnings are not necessarily comparable across firms, since accounting principles differ across countries and frequently entail some choice and scope.

- Earnings are not adjusted for the time value of money, ignoring anticipated inflation rates.

- A single year's earnings level does not indicate volatility over periods due to external influences and the firm's management decisions, thus, not reflecting the firm's business risk and financial risk appropriately.

- Maximization of earnings' implies to avoid dividend payouts, while dividend policy should consider investors' alternate return opportunities.

- Earnings account for the cost of debt and preferred stock, but omit opportunity costs for equity capital, thus, overstating value addition.

- Earnings-based return measures imply additional deficiencies:

- 'Return on sales', defined as net earnings divided by sales, does not reflect the capital used to generate earnings.

- 'Return on equity', defined as net earnings divided by equity, improves with an increasing share of debt capital, independent from the firm's value creation.

- After-interest 'return on net assets', defined as net earnings divided by total assets, charges for debt capital while omitting charges for equity capital.

- Before-interest 'return on net assets', defined as net earnings and after-tax interest expense divided by total assets, overcomes inconsistencies from prior metrics but omits any financing costs; further, respective divisional performance incentives do not necessarily maximize firm-wide returns.

To conclude, traditional accounting measures of firm performance are inappropriate VBM measures.

2.2.2 Introduction to Common Value-Based Performance Measures

Bias of traditional accounting measures suggests introduction of value-based accounting measures, to periodically review the firm's creation of shareholder value. Value-based metrics use either primarily cash flows or earnings after the cost of equity to determine value creation of management, as described in the following two sections. Common value-based performance measures derived from cash flows are:

- Discounted Cash Flow (DCF)

- Shareholder Value Added (SVA)

- Total Business Return (TBR)

Following valuation theory, the DCF model determines market value of the firm and, thus, shareholder value. Therefore, this concept builds the fundament of value-based performance measurement. Alternative value-based performance measures are required to reconcile with the discounted cash flow model, to qualify for VBM.

Derived from the DCF model, the single-period cash flow measures SVA and TBR avoid complexity and subjectivity of future cash flow estimates. SVA assumes current performance improvements to be infinitive. Contrarily, the TBR measure avoids any assumptions on future cash flows.

However, cash flow only provides accurate measures of value creation if it is considered over the life of the business or investment. Single-period cash flow fails to capture future value created by investments that earn above the cost of capital but involve substantial initial net investments, as seized by Stewart:[116] *"Abandon Cash Flow! However important cash flow may be as a measure of value, it is virtually useless as a measure of performance."*

Other than DCF analysis or single-period cash flow, 'Residual Income' (RI) measures periodic performance in terms of value creation, i.e. positive (negative) residual income implies that management added to (destroyed) shareholder value in the specific period.[117] Among the various residual income metrics that are derived from earnings instead of cash flow, the most commonly known are:

- Residual Income (RI)

- Economic Value Added (EVA)

- Economic Profit (EP)

- Cash Value Added (CVA)

The basic RI metric differs from net earnings by deducting charges for equity capital. The EVA and EP metric both represent residual income variants adjusted for accounting distortions. The CVA metric additionally integrates depreciation and inflation adjustments.

Less common residual income variants include the following:[118] 'Earnings less Risk-free Interest Charge' (E_RIC), [119] 'Refined Economic Value Added' (REVA), [120] 'Economic Margin' (EM), [121] 'Investment Recovery and Value Added' (IRVA), [122] 'Insurance Performance Measure' (IPM).[123]

In the following, the most frequently used measures, as listed above, are introduced.

---

[116] See Stewart (1991, p. 3).

[117] *"Economic profit measures the value created in a company in a single period of time"* (Copeland et al., 1994, p. 145).

[118] The review of residual income methods is not intended to be comprehensive in literature nor comprehensive in existing variations.

[119] See KPMG (2003), Kunz, Pfeiffer, and Schneider (2007), and Velthius (2004). E_RIC is promoted by auditing and consulting firm KPMG. It distinctively includes a capital charge based on a risk-free cost of capital.

[120] See Bacidore et al. (1997). REVA is an EVA variant applying the market rather than book values of capital for calculating the capital charge.

[121] See Obrycki and Resendes (2000). EM, marketed by the U.S. investment research firm Applied Finance Group, combines measurement methodologies from the EVA and CFROI concept.

[122] See Vélez-Pareja (2001). IRVA, developed by the academic researcher Vélez-Pareja, is a residual income variant based on real free cash flows.

[123] See Calandro and Lane (2002). IPM is another measure specifically developed for insurance firms.

Discounted Cash Flow (DCF)

The DCF model is based on free cash flows that represent cash generated through the firm's operations and investments, thus, available to all owners of the firm.[124] Accordingly, expected free cash flows of a planning period and a residual period are discounted by the weighted average cost of employed capital, *WACC*, to approximate total corporate value.[125] Then, corporate value, *CV*, is equal to the present value of future free cash flows, *FCF*, as follows:[126]

$$CV_0 = \sum_{t=1}^{\infty} \frac{FCF_t}{1 + WACC_t} .$$

DCF valuation, requiring estimates of future free cash flows, is rather complex and, above all, based on expectations rather than on actual accounted results. Nevertheless, the method is often used by several firms and corporations for evaluating the value creation associated to specific projects or within the planning process in order to assess the impact of periodical updates of long term plans, assigning firm value to existing assets and new strategies, based on free cash flows and a series of value drivers.[127]

Shareholder Value Added (SVA)

SVA is a variant of the DCF model (again, using the weighted average cost of capital, *WACC*, as discount rate) that assumes current performance improvements to be infinitely constant. Accordingly, the firm's periodic value creation equals the present value of infinite incremental operating profits after taxes after depreciation but before new investments, *OPAT*, less the present value of investments in fixed and working capital after depreciation, referred to as net investments, *NI*, notationally:[128]

$$SVA_{t-1} = \frac{OPAT_t - OPAT_{t-1}}{WACC} - \frac{NI_t}{(1 + WACC)} .$$

---

[124] Following cash flow statement classifications, free cash flows equal financing cash flows, also referred to as investors' cash flows. Free cash flow computation is carried out deducting cash taxes and net investments in net working capital, fixed assets, and other long-term assets from earnings before interest, taxes, depreciation and amortization.

[125] The FCF model follows standard valuation theory; however, in a VBM context, Rappaport (1986, pp. 50-77) introduces the 'Shareholder Value Approach' and Copeland et al. (1994, pp. 69-74) introduce the 'Entity DCF Model', both representing traditional FCF models.

[126] The DCF may also explicitly add the value of non-operating assets (such as marketable securities) and deduct future claims of other investors and stakeholders (such as preferred stockholders, debtholders and lessors) to compute corporate value.

[127] Rappaport shareholder value research for L.E.K. Consulting, see L.E.K. Consulting (1998a, 1998b).

[128] For definition, see Rappaport (1986, pp. 53, 72); the term 'Shareholder Value Added' has been introduced in the second edition of the initial publication (Rappaport, 1998). The formula can be rewritten to measure value added at the end of the period, as follows:

$$SVA_t = \frac{OPAT_t - OPAT_{t-1}}{WACC} \cdot (1 + WACC) - NI_t .$$

SVA stands out by integrating a built-in perpetuity of future earnings increases. The perpetuity avoids complexity of estimating future profits, simplifying implementation of a SVA-based performance measurement system. However, this simplification involves some disadvantages:[129] Above all, the perpetuity does not account for the fact that incremental profits are limited to a certain time window. Furthermore, especially in a cyclical business environment, the perpetuity makes the measure extremely volatile. Finally, the perpetuity leverages small changes in earnings increments. Consequently, it provides incentives for short-term earnings management, which disqualifies SVA as VBM measure.

Total Business Return (TBR)

Marketed by the Boston Consulting Group, TBR integrates the returns for existing assets and future asset growth by adding the return from free cash flows, FCF, (expressed as percentage of the firm's market value) to the percentage change in firm's market value, MV, notationally: [130]

$$TBR_t = \frac{FCF_t}{MV_{t-1}} + \frac{MV_t - MV_{t-1}}{MV_{t-1}}.$$

The change in market value considers any value created by management that is attributable to future plans. Therefore, TBR avoids estimates of individual cash flows, as required by the discounted cash flow model, and evades the assumption on infinitely constant performance improvements, implied by the SVA concept. However, TBR omits charges for the cost of capital, which the previous measures incorporated in the discount rate. Moreover, TBR calculation is based on market data, thus, questioning its validity for internal performance management.

---

[129] See Crasselt (2001).

[130] See Boston Consulting Group (2000a) and Olsen (2003); whereby, TBR is equivalent to the basic DCF one-year return for an investment (see Rappaport, 1986, p. 32); TBR may be alternatively measured over multiple (infinite) periods as discount rate setting expected free cash flows equal to beginning market value.

Residual Income (RI)

Contrary to net earnings, RI includes charges for equity capital. Commonly, RI is defined as net earnings before interest expenses, NE^{BI}, less charges for the cost of debt and equity capital, which are estimated based on the weighted average of cost of capital, $WACC$, and the book value of net assets, NA, as follows:

$$RI_t = NE_t^{BI} - WACC_t \cdot NA_{t-1} .$$

Moreover, RI can be rewritten as performance spread on net assets, using the profitability measure 'Return on Net Assets', $RONA$:[131]

$$RI_t = \left(\frac{NE_t^{BI}}{NA_{t-1}} - WACC_t \right) \cdot NA_{t-1} = \left(RONA_t - WACC_t \right) \cdot NA_{t-1}$$

Most important, RI reconciles with firm value as derived from the DCF model, thus, being consistent to the overall VBM objective to pursue shareholder value. The equivalence of the sum of future RIs discounted at a risk-adjusted rate, r, and net assets, NA, and corporate value, CV, is expressed as:[132]

$$CV_t = NA_t + \sum_{j=t+1}^{\infty} \frac{RI_j}{(1 + r_j)^j} .$$

In practice, basic RI, reflecting primarily single-period accounting earnings, is subject to management's earnings manipulation and penalizes for long-term investments. Therefore, consultancies developed residual income variants and corresponding return measures, incorporating a wide series of accounting adjustments, to reflect sustainable long-term performance.

Economic Value Added (EVA)

The consultancy Stern Stewart & Co. developed the EVA metric, an earnings-based residual income measure. EVA represents *"operating profits less the cost of all of the capital employed to produce those earnings"*.[133] EVA is defined as residual income of net operating profits after (adjusted) taxes, $NOPAT$, after covering the cost of capital, i.e. the weighted average cost of capital, $WACC$, times the beginning employed capital, $Capital$:[134]

$$EVA_t = NOPAT_t - WACC_t \cdot Capital_{t-1},$$

---

[131] The term RONA is yet commonly used, referring to net earnings over net capital. However, in the following, RONA is always referred to before-interest RONA, using net earnings before (not after) interest expenses, to make RONA comparable to the WACC.

[132] See O'Keefe and Lundholm (2001), Penman (1998), and Tham (2001b). For a discussion on the underlying assumptions, namely the clean surplus relation, see section 4.2.2.

[133] See Stewart (1991, p. 2).

[134] See Stewart (1991, pp. 136-138; 431-442; 742-744) for definition of EVA.

Just as RI, EVA can be rewritten where as spread between the return on capital employed, *ROCE*, and the cost of capital multiplied by the adjusted capital, notationally:[135]

$$EVA_t = \left(\frac{NOPAT_t}{Capital_{t-1}} - WACC_t \right) \cdot Capital_{t-1} = (ROCE_t - WACC_t) \cdot Capital_{t-1}.$$

Definition of NOPAT and Capital outlines that EVA includes several adjustments for accounting distortions via so-called 'equity equivalents'. [136] NOPAT principally equals net operating profits, the increase in bad debt reserve, Last-in-First-out (LIFO) reserves, and net capitalized research and development (R&D) expenses, the amortization of goodwill, and other operating income less cash operating taxes. Capital mainly consists of net assets, and various additions and deductions, as in particular the present value of non-capitalized operating leases, bad debt reserves, LIFO reserves, the cumulative goodwill amortization, capitalized R&D expenses, and cumulative unusual losses after taxes, excluding non-interest-bearing current liabilities, marketable securities and construction in progress.

The WACC estimate is based on the after-tax market cost of debt, the opportunity cost of equity and capital weights. The cost of debt equals the prevailing yield to maturity on the firm's outstanding traded debt. [137] The cost of equity is derived via the 'Capital Asset Pricing Model' model based on a risk-free rate from long-term government bonds, a market risk premium of 6% and a beta from regressing monthly common stock returns of the firm against monthly returns of the S&P 500 for five precedent years. Finally, capital weights reflect the target capital structure at market values over the trailing three years.

While ROCE provides a value-based substitute for the traditional return measures, it does not directly provide information on shareholder value added. Positive ROCE may imply management's addition to or destruction of shareholder value. Therefore, EVA, which integrates a minimum required return for debt and equity capital, provides superior information to shareholders. Then, ROCE higher or lower than the cost of capital implies additions to or subtractions from shareholder value, respectively.

---

[135] See Stewart (1991, pp. 85-93; 742-744) for definition of ROCE; while Stewart uses generic terms such as rate of return or return on total capital, the term 'Return on Capital Employed' facilitates to differentiate this return measure. The corresponding return ratio is also called rate of return on capital r and Stewart's r. To distinguish the EVA return ratio in this study, it is consistently referred to as ROCE.

[136] A more comprehensive approach to compute NOPAT and Capital is provided in section 2.3.1.

[137] If bond ratings of the firm's debt are not available, the cost of debt can be approximated by the rate paid by peers with equivalent bond rating based on the Stern Stewart Bond Rating Scoring System and the Moody's Bond Record or the Standard & Poor's Bond Guide. Anderson, Bey, and Weaver (2004) find that there is no material difference between EVAs using debt ratings or debt scorings.

Economic Profit (EP)

The consultancy McKinsey & Co. advocates the EP metric, which *"measures the dollars of economic value created by a company in a single year"*.[138] EP is either calculated as residual income of net operating profits less adjusted taxes, *NOPLAT*, after covering costs of capital or as spread between return on invested capital, *ROIC*, and the weighted average cost of capital, *WACC*, multiplied by the beginning invested capital, *IC*, notationally: [139]

$$EP_t = NOPLAT_t - WACC_t \cdot IC_{t-1} = \left(\frac{NOPLAT_t}{IC_{t-1}} - WACC \right)_t \cdot IC_{t-1} = (ROIC_t - WACC_t) \cdot IC_{t-1}.$$

Just as NOPAT and Capital (defined within the EVA concept), NOPLAT and IC include a set of accounting adjustments. NOPLAT is defined as pre-tax operating profits less cash taxes, i.e. operating earnings before interest and taxes less changes in deferred taxes and taxes adjusted for the tax shield on interest expense and taxes on interest and non-operating income; alternatively, NOPLAT can defined based on operating free cash flows.[140] IC represents the investments in the operations of the firm and equals the sum of operating working capital, net property, plant, and equipment, and other assets excluding non-interest-bearing current liabilities. Consequently, EP largely equals EVA, while possibly implying slight differences in the number or execution of accounting adjustments and the cost of capital computation. EP adjustments via so-called 'quasi equity' items are mostly equivalent but less wide-ranging than EVA adjustments.[141]

WACC computation involves the following components: The after-tax cost of debt equals the current expected yield to maturity in terms of the discounted cash flow return, which is also applied to leases. The cost of preferred stock equals promised dividends divided by market price. The cost of equity is estimated via the 'Capital Asset Pricing Model' (with a risk-free rate from T-bills, 10-year T-bonds or 30-year T-bonds, a 5% to 6% market risk premium, and a beta provided by BARRA) or the

---

[138] Copeland et al. (1994, p. 173); for a more detailed description, see Copeland et al. (1994, pp. 155-200) , which present EP as valuation model equivalent to the DCF approach; although the above formula uses beginning invested capital, capital may equivalently be measured as arithmethic average beginning and end-of-year balance (Copeland et al., 1994, p. 163); notably, the first edition of the 'Valuation' book by Copeland et al. did not include the Economic Profit approach.

[139] EP is introduced as single-period performance measure (Copeland et al., 1994, pp. 173; 239-273) as well as a multiple-period valuation model (Copeland et al., 1994, pp. 145-148).

[140] Then, NOPLAT equals free cash flows and gross investments less depreciation, see Copeland et al. (1994, p. 169).

[141] E.g., the authors avoid to evaluate assets at replacement costs *"for the simple reason that assets do not have to and may never be replaced"* (Copeland et al., 1994, p. 164); further, they only recommend valuation at market values *"when the realizable market value of the asset substantially exceeds the historical cost of book value"* (Copeland et al., 1994, p. 165).

multifactor 'Arbitrage Pricing Model'.[142] The target capital structure is based on a combination of current market values, the capital structure of peers and the firm's financing approach.

Just as ROCE, definition of ROIC does not incorporate a comparison to a minimum return required by investors and, thus, by itself does not provide information on value creation, that can be derived from its spread with respect to the WACC. On the contrary, EP, building on the previously defined ROIC measure, provides a measure of the firm's value addition.

Cash Value Added (CVA)

The Boston Consulting Group promotes the CVA metric, a cash flow-based residual income measure adjusted for inflation and depreciation.[143] CVA is either defined as residual income of sustainable cash flows, i.e. operating gross cash flows, *OGCF*, less economic depreciation, *ED*, after capital charges for gross investments, *GI*, or as spread between the cash flow return on investment, *CFROI*, and the weighted average cost of capital, *WACC*, multiplying by GI, notationally:

$$CVA_t = OGCF_t - ED_t - WACC_t \cdot GI_{t-1} = (CFROI - WACC_t) \cdot GI_{t-1}.$$

CFROI is determined as internal rate of return, including consideration of non-depreciating assets, *NDA*, fulfilling the following equation:

$$GI_t = \sum_{j=1}^{N} \frac{OGCF_j}{(1+CFROI)^j} + \frac{NDA_t}{(1+CFROI)^N}.$$

Alternatively, CFROI can be approximated as follows: [144]

$$CFROI_t = \frac{OGCF_t - ED_t^{CVA}}{GI_t}.$$

In the same way as EVA and EP, CVA implies several corrections for accounting distortions: Inflation-adjusted OGCF equals net earnings before minority interest, extraordinary items, and discontinued operations, and depreciation, after-tax interest

---

[142] However, the authors recognize that both the CAPM model and the APT model have some deficiencies: *"Both approaches have problems with their application. But they are theoretically correct: they are risk-adjusted and account for expected inflation"* (Copeland et al., 1994, p. 257).

[143] See Boston Consulting Group (1994, 1999). An alternate CVA metric with respect to the commonly known CVA proposed by Boston Consulting Group has been advertised by two Swedish business consultancies, distinguishes between non-strategic and strategic investments, and aims to manage several strategic investments, see Ottosson and Weissenrieder (1996) and Weissenrieder (1997).

[144] The single-period proxy formula only provides the same results as the 'Internal Rate of Return' (IRR) formula if the reinvestment rate (used for calculating the economic depreciation) is equal to the investment's internal rate of return, i.e., if the net present value of all investments equals zero. Whe the reinvestment rate is lower (higher) than the CFROI following the IRR formula, CFROI based on the single-period formula is lower (higher) than CFROI following the IRR formula. While HOLT Value Associates promote the IRR method, Boston Consulting Group suggests the single-period approximation formula.

expense on debt and leases, monetary holding gains, and FIFO profit. Contrary to FCF, OGCF includes an inflation adjustment, the net investment, and depreciation and amortization expenses. Inflation-adjusted GI include accumulated depreciation and amortization, Last-in-First-out (LIFO) reserves, non-capitalized operating leases, and a current dollar adjustment to gross plant and deferred taxes. Consequently, CVA stands out by additionally recognizing inflation rates. The inflation adjustment represents a key characteristic of the measure, however, adding complexity and potential for error. Moreover, research provides ambiguous results on whether the inflation adjustment increases accuracy.[145]

A method to estimate the WACC is not particularly emphasized.

The CFROI metric, marketed as stand-alone measure by HOLT Value Associates, represents an inflation-adjusted internal rate of return based on cash flows. [146] It is determined by setting expected OGCF and the final release of non-depreciating assets, *NDA*, equal to beginning GI, assuming assets in place generate constant real cash flows over the assets' anticipated economic life, *N*. OGCF and GI were yet introduced above. Inflation-adjusted NDA consist of net monetary assets, inventories, LIFO reserve, and inflation-adjusted land and improvements. N is approximated by dividing depreciating gross plant and equipment by depreciation expense, unless the specific lives of depreciating assets are provided. Contrary to the single-period approximation of CFROI, the internal rate of return CFROI spans multiple periods and, consequently, reflects the average returns of all existing projects at a certain period. Just as any of the previous return measures, the CFROI measure does not integrate a hurdle rate. Consequently, it has to be compared to an inflation-adjusted cost of capital in terms of CVA to determine whether the firm created value for its shareholders.

---

[145] Examining the impact of inflation on value-based performance measures, several authors expressed concerns that the nominal historical cost formulation of the Residual Income Valuation Relationship (RIVR) misvalues firms (see Bradley & Jarrell, 2008; Lee et al., 1999; Ritter & Warr, 2002). While one study testing two inflation-adjusted formulations of the RIVR finds that both are equivalent to the standard historical cost RIVR (O'Hanlon & Peasnell, 2004), another study examining the sensitivity of EVA to inflation shows significant inflation-induced distortions (Warr, 2005).

[146] See Madden (1998, 1999).

2.2.3 Evaluation of Metrics Favoring Residual Income Variants

The adoption of an effective performance metric may have a significant impact in determining the firm's decision-making, and, thus, shareholder value. However, metrics selected by management are commonly inappropriate: [147] *"Performance measures are one of the most misused management tools in business. Poorly chosen performance measures routinely create the wrong signals for managers, leading to poor decisions and subpar results."* Consequently, it is particularly important that the firm chooses an appropriate performance metric.

In the case of VBM, the fulfillment of six requirements constitutes a basic reference for evaluating the suitability of a performance measure:[148]

Foremost, the 'validity' criterion implies that the measurement method aims to capture the firm's value addition to shareholders and is consistent with an applicable value theory (i.e., it ideally reconciles with the market value of the firm over time). Equally important, the 'controllability' criterion means that corporate managers shall have the ability to influence the result of the performance measure and implicitly means that the metric is equally valid to measure performance of the overall firm, its divisions, its units, and its competitors. Third, the 'consistency' criterion relates to the measure's components that shall be consistently defined. Fourth, the 'objectiveness' criterion involves that performance measures are computed unambiguously from publicly readily available data, avoiding any subjective estimates. Fifth, the 'periodic delimitation' requirement implies that a performance measure shall reflect the performance of the respective investment period, independent from prior or succeeding periods. Finally, 'ease of implementation' refers to the simplicity of the measurement concept that ensures an effortless and proper implementation; additionally, this criterion ensures that the performance measure is communicable to employees, the board of directors, and investors.

Furthermore, considering the negative influence of distortions from book depreciation derived from the VBM experiences, another important criterion has to be added to the previous generic criteria and shall be considered in this study: 'Depreciation neutrality' prescribes that metrics shall either exclude or adjust book depreciation, to overcome any bias from accounting depreciation schedules.

As summarized by Figure 6, those seven criteria serve for an assessment of previously introduced performance measures, as follows:

- Validity: Traditional earnings measures are invalid VBM measures, since they are not directly related to firm value. Contrary to residual income measures, earnings measures do not account for costs of equity and, therefore, do not fulfill the

---

[147] See Knight (1998, p. 173).
[148] Several authors provide a wide range of factors for evaluating performance measures (see Knight, 1998, pp. 195-208; Madden, 1999, pp. 202-213; Rappaport, 1986, pp. 175-176; Schüler, 1998, pp. 18-85). I limit criteria to those that avoid subjective, qualitative assessments. Further, value-relevance of a measure, referring to its association with stock prices, is excluded, since driven by common practice and the availability of data.

valuation relationship. All cash flow and residual income measures described in the above section 2.2.2 are valid for value management, since they reconcile with market value of the firm over time and, therefore, explain creation of shareholder value. Furthermore, they indicate period by period the amount of shareholder value added or destroyed. Return measures derived from valid value metrics, such as ROCE, ROIC, and CFROI, partially fulfill the validity criterion since they lack integration of a hurdle rate to specify negative or positive value creation.

- Controllability: To a certain extent, firm managers can control the results from all earnings, cash flow, or residual income measures, except TBR. The TBR measure includes the market value of the firm, which is set by capital markets and, thus, not directly controllable by management.

- Consistency: As yet outlined before, traditional return measures commonly imply inconsistencies.[149] For instance, the return on equity metric neglects that also debt investments were crucial to generate returns; traditional after-tax return on net assets considers debt and equity capital in the denominator while only reflecting the cost of debt capital in the numerator. All other examined performance measures are consistently defined.

- Objectiveness: All measures, except the DCF method and the CVA approach, are computed with objectively defined data. The DCF method requires subjective estimates of future cash flows. The CVA and CFROI measure include subjective estimates of inflation.

- Periodic delimitation: All measures – except SVA – represent periodic performance measures, reflecting only the performance of the period to be examined. While the discounted cash flow and CFROI method both include future cash flows, these cash flows refer to operations and investments from the period to be examined. Equally, the change in market value included in the TBR measure relates to the firm performance of the period to be examined. On the contrary, the SVA measure assumes that current performance improvements endure infinitely.

- Ease of implementation: All measures – except CVA – are comprehensive, thus, excluding implementation or communication barriers. The CVA concept increases the accuracy of quantifications of shareholder value added via inflation adjustments. However, predicting inflation rates introduces considerable complexity to users such as operating managers and investors and involves considerable costs for assembling necessary information.[150]

- Depreciation neutrality: All cash flow measures, including the cash-based residual income measure CVA, are not affected by accounting depreciation. All traditional and earnings-based residual income measures deduct book depreciation and, therefore, imply some bias in their performance schedule.

---

[149] See section 2.2.1.
[150] See Madden (1999, pp. 203, 210-211).

	Traditional measures			Cash flow measures			Residual income measures			
	EPS	ROE	RONA	DCF	SVA	TBR	RI	EVA	EP	CVA
Validity	-	-	-	+	+	+	+	+	+	+
Controllability	+	+	+	+	+	-	+	+	+	+
Consistency	N/A	-	-	+	+	+	+	+	+	+
Objectiveness	+	+	+	-	+	+	+	+	+	-
Periodic delimitation	+	+	+	+	-	+	+	+	+	+
Ease of implementation	+	+	+	+	+	+	+	+	+	-
Depreciation neutrality	-	-	-	+	+	+	-	-	-	+

Figure 6   Qualitative Characteristics of Common Performance Metrics

All in all, traditional measures are all through unsuitable for VBM, since they are invalid, inconsistently defined, and influenced by book depreciation. Common cash flow metrics are inappropriate for varying reasons. To conclude, only the residual income measures RI, EVA, and EP fulfill all six universal VBM criteria. However, they are all affected by book depreciation.

Consequently, the qualitative review of common VBM metrics provides a first indication for the need to develop new residual income variants adjusted for accounting depreciation.

2.2.4 Widespread Familiarity of Managers with Value-Based Metrics

In the following, survey results outline the familiarity of financial managers with VBM and value-based performance measures. A high acquaintance of management with VBM provides some support for the relevance of the VBM concept and respective metrics.

First of all, there is evidence that VBM is well-known in the business community. E.g., a 1997 survey of 1,000 large U.S. industrial and non-financial service firms shows that corporate managers are mostly acquainted with VBM: 162 (87%) of 186 respondents have been familiar with VBM.[151]

Second, it seems that VBM is not only recognized but also commonly applied among corporations. For instance, a 2002 analysis of annual reports from large German DAX 30 firms finds that corporations frequently adopted VBM: 26 (87%) out of 30 firms disclosed VBM as overall management strategy.[152]

Third, there is evidence that EVA is the most commonly used value-based performance measure, if VBM is adopted. In the following the results of four studies are discussed in more detail.

To begin with, two studies on large German firms reveal that EVA and ROCE dominate, see Figure 7:[153]

A survey, carried out in 1998 analyzing 48 firms of the DAX 100, shows that EVA is by far leading: EVA (47%) prevails among applied performance measures, followed by the corresponding return measure ROCE (15%). Application of CVA (4%) and related CFROI (2%) is negligible. Traditional return measures, such as return on equity and return on investments are still relatively frequently used. Overall, residual income measures (51%) are as common as return measures (45%).

Four years later, in 2002, another survey of 35 DAX 30 firms finds evidence for a slight shift towards return measures, in particular towards ROCE: Return measures (63%) are somewhat more frequent than residual income measures (37%). ROCE (22%) dominates among used performance measures, followed by EVA (13%). Yet again, cash flow measures CVA (6%) and CFROI (3%) are quite rarely used. And once more, traditional return measures, such as return on investments and return on net assets, are still applied by some firms.

---

[151] See Ryan and Trahan (1999).
[152] See Fischer and Rödl (2003).
[153] See Fischer and Rödl (2003), examining 18 DAX 30 firms (of which 7 firms disclosed one measure, 8 firms disclose two measures and 4 firms disclosed three used measures), and Aders (2001), examining 48 DAX 100 firms.

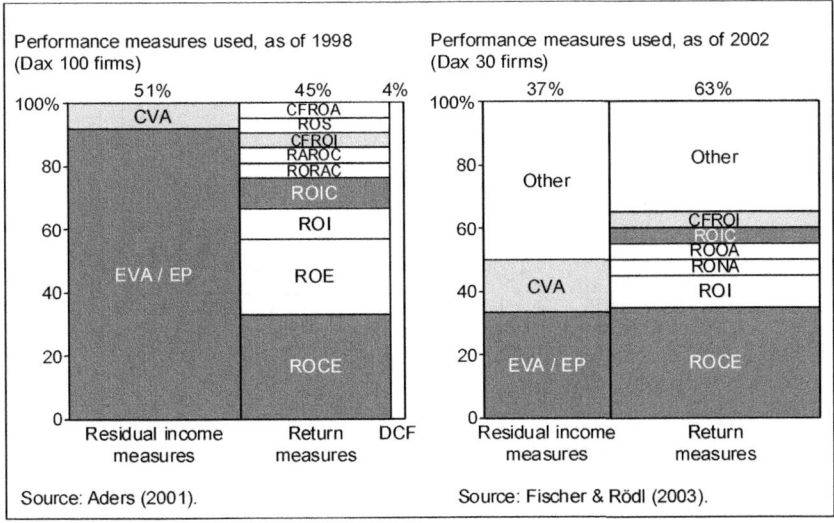

Figure 7   Value-Based Metrics Used by German VBM Firms

Moreover, two studies on U.S. firms underline the importance of EVA:

A 1998 survey sent to 1,300 U.S. firms clearly supports the predominance of EVA for U.S. firms:[154] Of 88 respondents, 37 firms (42%) adopted EVA, 32 firms (36%) only considered adopting EVA, 17 firms (19%) do not consider adopting EVA, and 2 firms (2%) adopted and subsequently dropped EVA. Consequently, EVA is commonly considered and the share of unsatisfied EVA users is quite low.

Furthermore, a 1997 survey of 162 chief financial officers of large U.S. industrial and non-financial service firms, being familiar with VBM, provides further evidence that EVA is most well-known among U.S. firms, see Figure 8:[155] While EVA is the most commonly known, however, DCF is most frequently applied, followed by EVA, CFROI, and ROIC. However, DCF users are relatively often unsatisfied. Other value-based performance measurement methods do only play a minor role.[156]

---

[154] See Dodd and Johns (1999)
[155] See Ryan and Trahan (1999).
[156] Including Amp Value Added, KPMG Critical Factors, IRR, Total Enterprise DCF, LBO Base, Towers Perrin Value Based Management, Cash Flow Return with Cost of Capital Threshold, Marakon DCF, SVA, and Braxton.

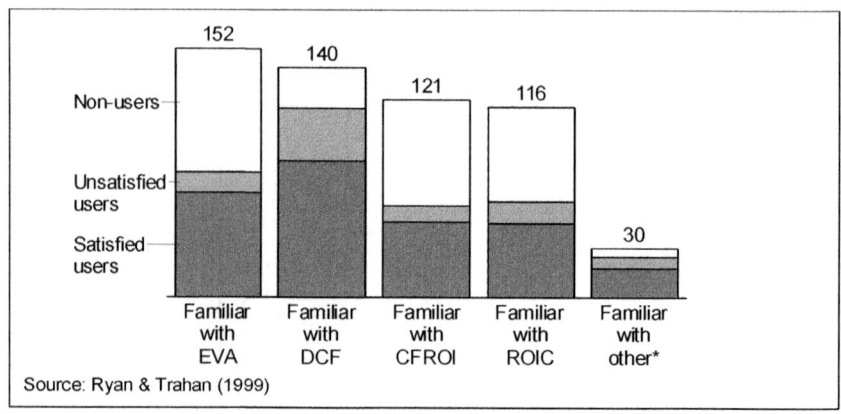

Figure 8   Value-Based Metrics Used by U.S. VBM Firms

Overall, results may be affected by response bias. However, there is some evidence from German and U.S. firms that VBM is commonly known and adopted. Further, results suggest that earnings-based EVA is frequently applied while cash-based CVA is rarely used. While the cash measure CVA avoids book depreciation, its complex calculation methodology may hinder corporate managers from its adoption.

2.2.5   Multiple Uses of Value-Based Metrics

The following overview of specific applications of value-based metrics provides rationales underlying their adoption. As shown in Table 1, value-based metrics can be employed in diverse corporate functions and business areas by various interest groups from the business and financial community.

Applications of Value-Based Metrics	User Groups
- Corporate management, including strategic and operating planning, supply chain management, mergers & acquisitions, and management compensation	- Top management - Corporate divisions - Business units
- Group and divisional management	- (Institutional) investors
- Controlling	- Board of directors
- Financial reporting	- Security analysts
- Internal communication	- Portfolio managers
- Investor relations	- Consultants for VBM
- Performance monitoring	- Business journalists
- Equity valuation	
- Portfolio selection	

Table 1   Common Uses of Value-Based Performance Measures

The five major applications of value-based performance measures are:
• Corporate management
• Cross-hierarchical performance management
• Internal and external communication
• Investment analysis
• League tables

Corporate Management
    First and foremost, VBM measures are to be applied ex ante for strategic and operating planning, i.e. for evaluation of internal investments and external growth; over the time, value-based decision-making has been extended from individual projects to strategic plans of strategic business units (SBUs) and the company itself.[157] Furthermore, there is wide agreement that tying management compensation to value-based performance measures is essential for the success of VBM.[158] Increasingly, VBM measures are also utilized to forecast, evaluate and manage mergers and

---

[157] See Rappaport (1986, pp. 2-3).
[158] Several authors regard the adoption of VBM measures for management compensation plans as essential for the success of a VBM system (see Knight, 1998, p. 215; Martin & Petty, 2000, p. 9; O'Byrne & Young, 2001, p. 111).

acquisitions, in order to avoid overpaying, to reevaluate acquisitions, and to meet expected post-acquisition performance targets. [159] For strategic planning, a VBM measure can be integrated into a 'Balanced Scorecard' as lagging indicator of value creation, to combine the strength of a VBM measure, focusing management on shareholder value creation, with the strength of the Balanced Scorecard, clarifying exhaustively underlying value drivers to support decision-making.[160] For supply chain management, a VBM metric may supplement cost information from an 'Activity Based Costing' system, to improve internal operations and to understand system-wide performance and contributions of each supplier.[161]

Results from a 1997 survey of 162 large U.S. firms, using value-based performance measures, reveals that DCF and CFROI are narrowly used, while EVA and ROIC are commonly applied across corporate functions, see Figure 9:[162]

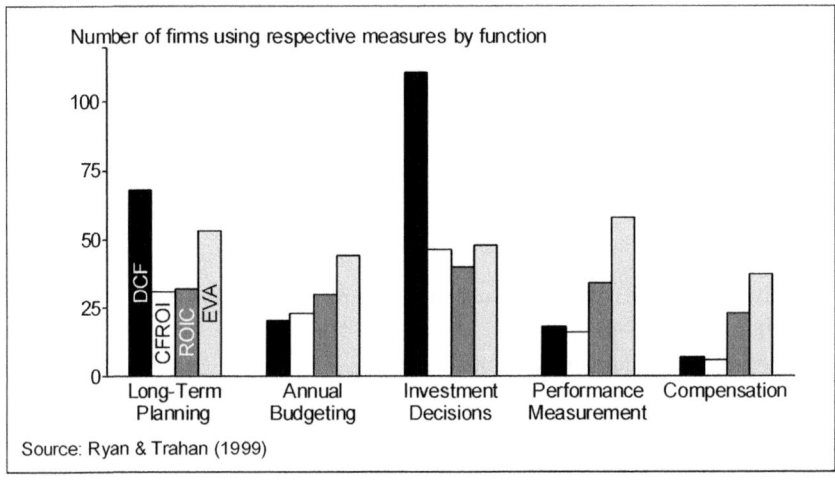

Figure 9   Usage of VBM Metrics by Corporate Area

---

[159] See Gandhok, Dwivedi, and Lal (2001), Sirower and O'Byrne (1998), and Yook (2004).

[160] The Balanced Scorecard represents a performance management tool integrating customer, internal process, innovation and financial perspective while balancing leading and lagged, financial and non-financial, and quantitative and qualitative performance drivers; research emphasizes the necessity of integrating BSC and the VBM metric (see Fletcher & Smith, 2004; O'Byrne & Young, 2001; Shinder & McDowell, 1999).

[161] Activity Based Costing represents a cost measurement system that provides a cost for each product, service or customer, analyzing direct and indirect activities needed for production and allocating associated costs; research outlines frameworks for integrating Activity Based Costing and VBM measures (see Pohlen & Coleman, 2005; Shinder & McDowell, 1999).

[162] See Ryan and Trahan (1999); seven responses referring to other applications including acquisitions, education, reference, pricing, and ad hoc use, not further discussed.

Of 112 DCF users, almost all (97%) apply DCF for investment decisions, conducting net present value project evaluations, and a majority (59%) use DCF for long-term planning. Likewise, 82% and 55% of 56 CFROI users employ the measure for investment decisions and long-term planning, respectively. Since utilization of DCF or CFROI in other areas is immaterial, both may be considered a measure for investment analysis rather than for VBM. Contrarily, 76 EVA users and 59 ROIC users apply those measures across diverse functions, including long-term planning, annual budgeting, investment decisions, performance measurement and compensation.

Moreover, another survey of 83 large U.S. firms that use value-based performance measures for human resources management demonstrates that value-based performance measures most commonly determine bonuses of Chief Executive Officers and other executives, see Figure 10:[163]

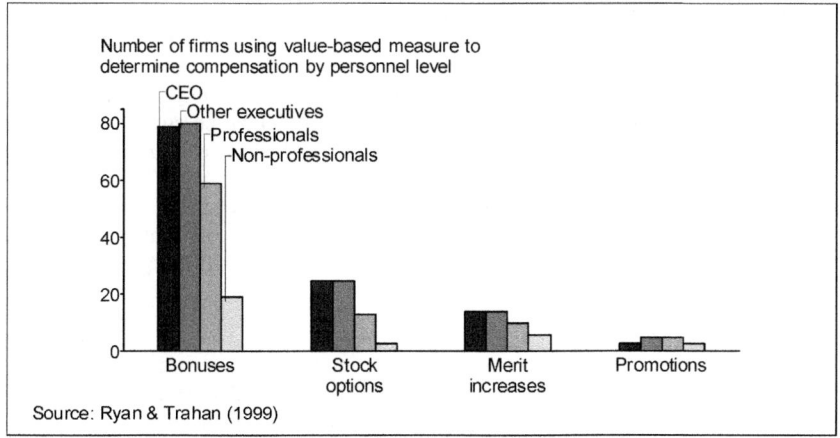

Figure 10   Usage of VBM Metrics for Compensation and Promotion

95% of surveyed firms use value-based bonuses for the Chief Executive Officers and other executives, while only 30% and 17% of firms provide stock options and value-based merit increases to executives, respectively. Value-based promotions are generally quite rare, used by about 5% of surveyed firms. While value-based bonuses for professional and non-professional employees are yet used by 70% and 20% of analyzed firms, respectively, other value-based compensation for employees is fairly unusual.

---

[163] See Ryan and Trahan (1999).

Cross-Hierarchical Performance Management

Ideally all hierarchical levels and organizational units apply value-based performance measures.[164] Therefore, VBM measures shall encompass corporate as well as divisional management. For successful divisional performance management, a diagnostic and prescriptive framework first supports management in segmenting the business operations in practicable performance centers.[165] Then, a value-based performance measure serves to consistently manage and reward divisional performance in the interest of shareholders.[166] Likewise, VBM measures apply to manage subsidiaries, and single plants and offices.[167] Going further, value-based measures can be even applied to measure value added by customer.[168]

Internal and External Communication

The firm's internal and external communication represents another important application of a VBM metric.[169] Internal communication of value-based strategic goals, key management processes, and operating decision-making is essential for employees to understand value creation and their impact on it. Equally important, communication of value creation to the investment community supports capital markets to recognize the firm's value addition. Finally, VBM measures may improve performance monitoring from a value-based view and, thus, corporate governance mechanisms available for the board of directors.[170]

Empirical evidence underlines the importance of investor relations: The stock price of firms that have adopted VBM and communicated superior value creation to investors increased by 21% p.a. over the period December 1990-1996, exceeding the 16% p.a. value growth of the S&P 500.[171]

Investment Analysis

Security analysts and portfolio managers can use value-based metrics for value-based stock valuation and portfolio selection.[172] Therefore, value-based measures have yet been adopted by investment firms, including buy-side firms such as GAM and Oppenheimer Capital and sell-side firms such as Credit Suisse First Boston, Deutsche Bank Securities, ADP and Goldman Sachs, for stock selection, portfolio construction,

---

[164] See section 2.1.2.

[165] See Hodak (2000).

[166] See Jensen and Meckling (1990) and Martin and Petty (2000, p. 45).

[167] E.g., Wills Group's CEO Lock Wills comments on successful implementation of a VBM system based on EVA across all levels: *"EVA allows us to measure at the parent level, the subsidiary level, and even down to the individual locations with a single measure, not a series of measures."* (Walbert, 1995, p. 109).

[168] RandMcNally's chief financial officer James Habschmidt states: *"In our book business, they looked at EVA by customer and have seen that impact in payment term extensions."* (Walbert, 1995, p. 109).

[169] See Knight (1998, pp. 269-272) and Martin and Petty (2000, p. 9).

[170] See Straw, Peck, and Keller (2000)

[171] See Knight (1998, p. 271).

[172] See Abate and Grant (2001, pp. 13-66) and Wolin and Klopukh (2000).

and risk control processes; based on the rationale that equity style investing inadequately reflects the value creation of the firms, investment firms introduced value-based investment models (for instance, Credit Suisse First Boston, HOLT Value Associates, and Deutsche Bank Securities using an economic profit based stock evaluation model) and consultancies introduced stock rating systems for investment managers (such as the Performance Risk Valuation Investment Technology, PRVit, model by Stern Stewart).[173] Notably, analysts also adopted a value-based approach to evaluate performance of different parts of a firm such as divisions or geographic markets, using public data and information individually requested from firms and their suppliers and industry peers.[174]

League Tables

The rise of VBM resulted in value-based league tables, ranking firms by a measure of shareholder value creation. Value-based rankings imply several uses:[175] Foremost, rankings identify top and bottom creators of shareholder wealth. [176] Further, a comparison between the internal VBM metric, such as EVA, and the corresponding market measure, such as MVA, either serves to identify mispriced equity securities or to draw inferences on investor expectations on future operating performance. [177] Finally, time-series and cross-sectional analysis allows to detect trends and to make comparisons among firms and industry groups, and industry-wide capital efficiency judgments.

Apart from predestined uses, prevailing value-based league tables shape the cognition of VBM among business managers and the investment community. Further, researchers commonly apply ranking data to examine the value-relevance of value-based metrics.

---

[173] See Abate, Grant, and Stewart (2004), Herzberg (1998), and Jackson (1996); whereby, investment firms commonly referred to EVA as value-based measure, see e.g. the Credit Suisse First Boston's landmark 1996 Economic Value Added Conference or the Goldman Sachs' May 1997 conference titled 'EVA and Return on Capital: Roads to Shareholder Wealth' that introduced its EVA-based research platform.

[174] E.g., global equity research at Credit Suissse First Boston or Trainer, Wortham & Co., see Ross (1997, p. 119)

[175] See O'Byrne and Young (2001, pp. 8-13, 81-103) and Walbert (1994, pp. 110-111).

[176] This implies that wealth creators (destroyers) have been largely investing in projects with positive (negative) net present value.

[177] In the former case, overvalued (undervalued) stock shows relatively low (high) levels of economic profit, based on a comparison with firms of comparative market value. In the later case, an equity with comparatively high (low) levels of market value or low (high) levels of economic profit imply that investors expect considerable improvements (deteriorations) in the firm's future economic profits. Notably, investors are more likely to expect performance improvements, given that the contemporaneous relationship between MVA and EVA is more robust for wealth creators than for wealth destroyers.

Table 2 outlines chronologically seven recognized performance rankings that are publicly available via a well-known print publication,[178] report ranks for individual firms,[179] and are based on a measure of shareholder value addition.[180]

At first, academicians from the London Business School published a value-based league table in 1991, ranking over 2,000 firms – excluding financial services, insurance, and real estate firms – by 'Added Value' (AV).181 AV is defined as "the amount by which the value of corporate output exceeds the value of all the inputs which the company uses – including not only material, but also capital and labor".182 Notably, this AV ranking is exclusively based on an internal operative measure of value creation, without using any market data. While the ranking has not been annually continued, primarily business consultancies took on the design, and publication of league tables.

Most prominently, Stern Stewart & Co. initiated its Stern Stewart Performance 1000 in 1994, ranking the 1,000 largest U.S. firms – excluding electric utilities and real estate firms – by 'Market Value Added' (MVA).[183] MVA is defined as market value less book value of invested capital.[184] Based on market data, MVA explain how much shareholder wealth a firm created or destroyed. The ranking additionally provides the firm's EVA, profitability index and company type.[185] The annual ranking of U.S. firms

---

[178] Stern Stewart provides MVA analysis on more than 1,000 U.S. firms and other geographic markets than the U.S., e.g. performance rankings for 500 U.K. firms, 3,000 U.S. firms of the Russell index, and German, Italian, Spanish, Canadian, Australian and Indian firms. However, these rankings are for purchase only.

[179] E.g., Credit Suisse First Boston provide ROIC – WACC spreads by country, sector, company, and entity in an in-house publication of September 29, 1998, titled 'Frontiers of Strategy'. However, firm rankings are only illustratively shown for the computers industry. McKinsey presents a study titled 'Comparing performance when invested capital is low' in its in-house publication The McKinsey Quarterly (No. 17, Autumn 2005, pp. 17). However economic profit, the ranking variable, is only illustratively provided for three industry groups.

[180] The 2003 Credit Suisse First Boston in-house study 'Champions League der Konzerne' ranks firms from 16 industries by sales, also providing CFROI, defined as a sustainable current-dollar cash flow divided by current-dollar cash invested, and market expectations, explaining the stock price via the 5-year CFROI and current sales. While CFROI represents a measure of economic value addition, the league table is based on sales, a measure not related to shareholder wealth.

[181] See Davis, Flanders, and Star (1991).

[182] For a description of the 'Added Value' (AV) concept, see Davis and Kay (1990). Accordingly, equals 'Operating Profits' (OP) less a capital charge that equals 'Cost of Capital' (CoC) times beginning 'Capital Employed' (CE). CE includes land and property, fittings, and physical stocks used by the firm, preferably valued at current cost. The long-term interest rate on safe corporate bonds serves as CoC; notably, CoC does not reflect an average cost of debt and equity capital. To avoid size dependence, AV can be analyzed as intensity, i.e. as ratio of AV to the input costs in terms of labor and capital employed. However, definition of AV involved some weakness in its cost of capital estimate.

[183] See Ross (1996, 1997, 1998, 1999) and Walbert (1994, 1995).

[184] See section 2.3.3.

[185] The profitability index is calculated as 5-year average return on capital divided by 5-year average cost of capital. The company type is assigned to one of five predefined types according to the profitability spread and capital growth of the firm.

gained more attention once published by the Fortune magazine, starting in 1998:[186] It promoted MVA analysis suggesting that *"no investor should be without"*.[187] In 2000, the ranking was extended to cover 200 largest London Stock Exchange quoted firms.[188] In 2001, the consultancy switched to ranking firms by 'Wealth Added Index' (WAI).[189] Principally, the WAI measures shareholder value creation by deducting the opportunity cost of equity from total stock returns.[190] Notably, rankings provide evidence that top wealth creators remain stable over several years and bottom performance occurs mostly in capital-intensive firms.[191]

When Stern Stewart started its publications in 1994, L.E.K. Consulting commenced publishing league tables in the Wall Street Journal.[192] The listing, designed by VBM pioneer Rappaport, ranks the largest 1,000 firms in the Dow Jones U.S. Total Market Index by the total return to shareholders over one-, three-, five-, and ten-year holding periods.[193] However, the ranking omits information on value creation from an internal value-based metric.

Starting in 1999, Boston Consulting Group provided another annual performance report, ranking firms according to five-year average annual changes in total shareholder returns and, starting in 2001, according to 'Total Business Returns' (TBR).[194] In some editions, the report additionally provides CVA, an 'expectation premium', the relative importance of value drivers, and a decomposition of shareholder returns.[195] While the first editions explain shareholder returns via the internal CVA measure, the latest editions decompose returns.

---

[186] See Tully (1998). Following, the ranking was reduced to the 200 largest U.S. firms by market capitalization and supplemented by information on the 'Future Growth Value' (FGV). FGV, defined as market value less the perpetuity of the latest NOPAT discounted at the cost of capital, allows assessing whether the proportion of firm value is reasonable.

[187] See Colvin (2000, p. 210).

[188] See Stern (2000). The listing also provides the return on capital employed, the cost of capital and the FGV. Understanding FGV was claimed to be especially important for firms aiming to maximize shareholder value.

[189] See Anonymous ("Marked by the market," 2001, December).

[190] Notationally, WAI is defined as WAI = Δ market value + dividends – required return– shares issued; whereby, the required return is estimated via the CAPM model. Over time, the average company is supposed to have zero Wealth Added.

[191] See Ross (1998, p. 116, 1999, p. 122).

[192] See Anonymous ("How the Scoreboard Ranking Were Compiled," 2008) and Rappaport (2001).

[193] Total return includes price appreciation or depreciation, and any reinvestment from cash dividends, rights and warrant offerings, and cash equivalents, such as stock received in spin-offs, and is also adjusted for stock splits, stock dividends, and recapitalizations.

[194] Boston Consulting Group (1999, 2000b, 2001, 2002, 2003, 2004, 2005, 2006, 2007, 2008). For TBR definition, see section 2.2.2.

[195] CVA is defined as gross cash flow less economic depreciation and a capital charge. The 'Expectation Premium' is calculated as the firm's market capitalization and debt less the present value of additional cash flow due to growth and profitability and the current performance discounted to perpetuity.

Affiliation	Publication	Coverage	Ranking variable	Related VBM metric	Additional variables
London Business School	Busiess Sntrategy Review (1991)	Over 2,000 firms, worldwide	Added Value (AV)	N/A	- Added Value / Sales - Return on capital employed (ROCE)
Stern Stewart	Journal of Applied Corporate Finance (1994-99)	1,000 largest U.S. firms by market value	Market Value Added (MVA)	Economic Value Added (EVA)	- Profitability index - Company type
L.E.K. Consulting	Wall Street Journal (1994-2008)	1,000 largest firms of the U.S. Dow Jones Index	5-year stock return	N/A	- 1-year return - 3-year return - 10-year return
Boston Consulting Group	In-house publication (1999-2008)	Over 500 large quoted firms, worldwide	Total Business / Shareholder Return (TBR/TSR)	Cash Value Added (CVA)	- Expectation premium - Value drivers
Stern Stewart	Fortune magazine (1998-2000)	200 largest U.S. firms by market value	Market Value Added (MVA)	Economic Value Added (EVA)	- Future Growth Value (FGV) - Future Growth Value as % of market value
Stern Stewart	The Sunday Times (2000)	200 largest U.K. firms by MVA	Market Value Added (MVA)	Economic Value Added (EVA)	- Return on capital employed (ROCE) - Cost of capital - Future Growth Value
Stern Stewart	The Economist (2001)	5,069 largest firms by market value	Wealth Added Index (WAI)	N/A	- Cost of equity - Total shareholder return

Table 2   Outline of Value-Based Performance Rankings

2.2.6 VBM Systems Effectively Addressing Management Decisions

The adoption of VBM and, in particular, the implementation of value-based compensation schemes, aims to align management behavior with shareholder interests. Specifically, value-based residual income incentive schemes supposed to be superior to traditional earnings-based compensation plans in encouraging management to maximize shareholder wealth.[196]

From an operating perspective, RI incentives are supposed to increase profits and returns by using invested capital more efficiently, thus, increasing asset turnover. From an investing view, RI incentives may result in the re-allocation of capital towards activities that earn more than their capital costs. From a financing perspective, adoption of RI incentives should influence financing decisions, i.e. payouts of capital earning less than the cost of capital as tax-favored share repurchases rather than cash dividends. Altogether, changes in decision-making in line with VBM shall increase shareholder wealth.

Numerous researchers examined whether corporate managers respond to value-based (residual income based) incentives in ways that contribute to shareholder wealth, to evaluate the effectiveness of VBM adoption. While results vaguely indicate that firms outperform after VBM adoption, studies show clear evidence on changes in management's decision-making consistent with VBM objectives, as shown in Table 3.

Author (Year)	Sample	Post-adoption increase in performance	Altered decision-making as expected
Wallace (1997)	40 U.S. firms with value-based incentives	Yes	Yes
Kleiman (1999)	66 U.S. firms with EVA incentives	Yes	Yes
Cordeiro & Kent (2001)	63 U.S. firms with EVA incentives	No	N/A
Cahan et al. (2002)	52 New Zealand managers with EVA incentives	Yes	N/A
Prakash et al. (2003)	Up to 48 U.S. firms with EVA incentives	N/A	Yes
Griffith (2004)	63 U.S. firms with EVA incentives	No	N/A
Hogan & Lewis (2005)	108 firms with value-based incentives	Yes	Yes
Ryan & Trahan (2007)	84 firms with value-based incentives	Yes	Yes

Table 3   Evidence on the Association of VBM Adoption with Firm Performance

---

[196] For example, see Bowen and Wallace (1999).

However, results may be to some extent affected by non-random samples – commonly small samples of Stern Stewart clients – and changes in operations that are not related to VBM adoption, such as management realignments, strategic repositioning, and restructuring. In the following, studies are briefly outlined in chronological order.

Examining changes in management's operating, investing and financing decisions and overall performance in response to RI incentives, Wallace (1997) analyzes 40 adopters of RI incentives and matched-pair non-adopters, comparing a 5-year pre-adoption and 3-year post-adoption period. In fact, Wallace finds that asset turnover relatively increases by 14%, asset dispositions relative increase by 100%, new investments relatively decrease by 21%, share repurchases relatively increase by 112%, and cash dividends relatively increase by 1% for RI adopters after VBM adoption. Therefore, the study provides evidence that RI incentive plans are effective in altering management decisions. Additionally, it finds a considerable relative increase in residual income for firms adopting RI incentive plans.

Kleiman (1999) examines shareholder returns and management decisions of 66 EVA adopters relative to the closest-matched peer, comparing a 3-year pre-adoption and 4-year post-adoption period. He finds that EVA firms significantly outperform their competitors in the year of adoption and the 3-year post-adoption period, providing cumulative abnormal returns of 7.8%. Furthermore, EVA adopters continue to reduce the cash conversion cycle, increase their financial leverage and substantially improve operating income and operating margins after adoption of EVA incentives.[197] All in all, Kleiman validates prior results confirming the effectiveness of RI-based compensation plans.

Controlling for size, leverage, industry, earnings history and coverage, Cordeiro and Kent (2001) analyze the association of EVA adoption of 63 firms with earnings forecasts. However, they did not find a significant relationship between EVA adoption and forecasts of future earnings performance.[198] The lacking relationship may result from the fact that EVA-based incentives may hinder considerable long-term investments required for earnings growth. Namely, EVA relatively decreases due to up-front capital commitments.[199]

Studying the performance of mid-level managers, Cahan, Lal and Riceman (2002) examine the performance of 52 managers adopting EVA incentives relative to 65 managers continuing to use traditional earnings incentives. The authors find that managers with EVA incentives that understand EVA outperform. Further analysis

---

[197] Contrary to above findings, there is no evidence of a reduction of capital expenditures. However, both results are still in line with VBM objectives: Since new investments, even with the same margin than ongoing investments, result in a deterioration of EVA due to age bias, see section 3.1, EVA tends to imply a bias against new investments. However, if new investments earn the cost of capital and imply considerable positive EVA, they may even increase the overall EVA.

[198] EVA adoption explained 0.8%, 1.1%, and -3.8% of end-of-year, end-of-next-year, and next five-year earnings forecasts.

[199] See section 3.1.

suggests that performance increases are due to increased consistency in the evaluation reward process.

Combining adoption research with an analysis of firm risk, Prakash, Chang, Davidson, and Lee (2003a, 2003b) examine significant differences of key financial ratios and systematic and unsystematic risk, based on a 4-year post-adoption and 4-year pre-adoption period. As expected, liquidity ratios decreased, while asset turnover, profitability, market value ratios increased, and changes in financial ratios are all significant at the 0.01 level. Systematic market risk mostly decreased, possibly as result of market conditions and a decrease in financial leverage, since EVA adoption aims to reduce the cost of debt. Since unsystematic risk mostly increased, total risk increased after EVA adoption, in line with EVA aiming at increasing shareholders' returns.

Comparing firms that adopted the EVA incentive system only partially and completely, Griffith (2004) examined performance of 63 EVA adopters, of which 30 adopted the complete EVA incentive system including uncapped bonuses and bonus banks for extraordinary awards, in a 5-year post-adoption period relative to the Russell 3000. EVA firms significantly underperform with a cumulative abnormal loss of 55%. However, underperformance is primarily due to the significant loss of partial adopters of 115% after EVA adoption.

Examining the impact of VBM adoption on performance in combination with a classification of VBM adopters, Hogan and Lewis (2005) analyze the performance of 108 firms that adopted value-based compensation plans. First, the study shows significant improvements in operating performance and shareholder value following adoption. [200] Furthermore, classifying firms into 'anticipated adopters', 'surprise adopters', 'anticipated non-adopters', and 'surprise non-adopters', they find that 'anticipated adopters' outperform 'surprise non-adopters', possibly due to a more cautious assessment of the applicability to individual firm characteristics and greater implementation efforts.

Reviewing post-adoption economic performance and management behavior as well as factors determining the likelihood of VBM adoption, Ryan and Trahan (2007) examine 84 firms that introduced a VBM system relative to matched-firm-adjusted returns for a 5-year post-adoption period. Results reveal a significant improvement of RI divided by invested capital by 8% following the adoption of a VBM system. Additionally, findings show that firms reduce capital expenditures following VBM adoption, regardless of the firm's growth opportunities. Finally, the study reveals that the likelihood of adopting a value-based compensation plans is greater if they perform better prior to adoption or if other firms in the industry yet use value-based compensation plans.

All in all, results indicate that VBM is an effective management strategy to align management behavior with shareholder objectives and to maximize shareholder

---

[200] The median adopter's operating income before depreciation divided by total assets increases from 4% prior to adoption to 5% four years after adoption. The market-to-book ratio increases from 2.4 preceding the adoption to 2.8 four years following the adoption.

wealth. Moreover, adoption of a complete and consistent value-based incentive system, a thoroughly planned implementation, and well-trained employees seem to be critical for success.

2.3 The Economic Value Added (EVA®) Concept

The precious chapters yet pointed out the prevalence of the EVA measure. In the following, a review of its computation, its internal and external uses, and qualitative characteristics allows a profound understanding of the metric.

2.3.1 EVA Computation Going Far beyond Subtraction of Equity Costs

As yet mentioned in section 2.2.2, EVA is defined as net operating profits after the cost of debt and equity capital, notationally $EVA_t = NOPAT_t - WACC_t \cdot Capital_{t-1}$, where $NOPAT$ = (adjusted) net operating profit after taxes, $WACC$ = weighted average cost of capital, and $Capital$ = employed (adjusted) net capital.

Consequently, EVA is primarily coined by three elements: NOPAT, Capital, and WACC. NOPAT are *"profits derived from the company's operations after taxes but before financing costs and non-cash-bookkeeping entries.* [201] In general, periodical NOPAT is related to beginning capital, to conform to the residual income valuation relationship.[202] Capital includes the firm's operating assets and a series of adjustments, including 'equity equivalents' that capture value of assets beyond their reported book value. [203] The WACC is estimated based on the market cost of debt and equity, representing *"the minimum acceptable return on investment [...], a cutoff rate that must be earned in order to create value."*[204]

However, earnings and capital elements that determine EVA include several adjustments for accounting distortions. Stewart identified at least 164 potential adjustments in GAAP-based financial statements.[205]

As shown in Figure 11, EVA adjustments can be divided into conceptual adjustments, excluding non-operating and non-recurring items as well as any financing

---

[201] See Stewart (1991, p. 86).
[202] For the valuation relationship, see section 4.2.20. Notably, Stewart (1991, p. 742) suggests replacing beginning capital by the arithmetic average capital *"if assets declined by more than 20% over the year or if acquisition expenditures totaled more than 20% of average assets".*
[203] *"Equity Equivalents (EEs) gross up the standard accounting book value into [...] economic book value, which is a truer measure of the cash that investors have put at risk in the firm and upon they expect their returns to accrue"* (G. B. Stewart, 1991, p. 91); Stewart (1991, p. 742) suggests replacing beginning capital by the arithmetic average capital *"if assets declined by more than 20% over the year or if acquisition expenditures totaled more than 20% of average assets".*
[204] See Stewart (1991, p. 431).
[205] See Stewart (1994, p. 73).

costs, and US-GAAP related adjustments, introducing debt and equity equivalents to account for assets not recognized by US-GAAP.[206]

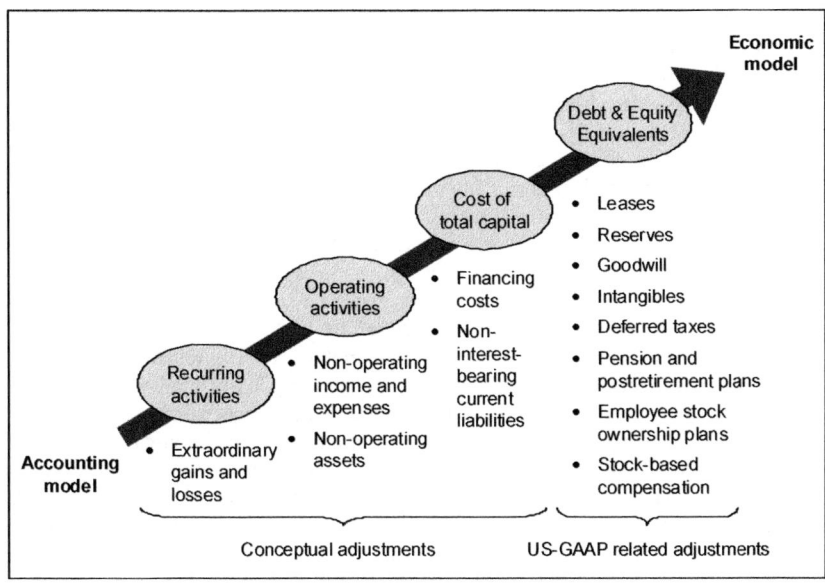

Figure 11   Key Characteristics of Economic Value Added (EVA)

First of all, EVA shall reflect recurring activities of the firm, which implies that current-period and cumulative after-tax unusual losses (gains) are added back to (subtracted from) earnings and capital, respectively.[207] Extraordinary items include, e.g., restructuring charges, gains on disposal of assets and non-cash write-down of impaired goodwill.

Second, the portion of capital producing operating profits shall correspond to operating profits, requiring excluding non-operative income (expenses) from earnings,[208] and non-operating assets from capital.[209] Usually, non-operative income equals interest income and other non-operating income and expenses. Common non-operating

---

[206] The presented classification aims to improve the understandability of extensive adjustments and presumably summarizes 164 potential adjustments, recognizing that EVA adjustments commonly have an impact on earnings and capital and often affect several accounting items. Stewart does not categorize accounting adjustments in a comprehensive manner and, divergently, he classifies the adjustment for unusual items as equity equivalent.

[207] See Stewart (1991, pp. 2, 116-117, 2003, pp. 76-78).

[208] See Stewart (1991, p. 86).

[209] See Stewart (1991, p. 2).

assets are marketable securities, excess cash, equity investments, capitalized own stock, construction in progress, real estate rented to a third party, apartments provided to employees, closed-down facilities and installations, assets acquired in the course of mergers and available for sale, and deferred debt issuance costs

Third, total costs for all sources of debt and equity capital are deducted from earnings via a capital charge on total capital. Consequently, EVA requires excluding any financing costs from earnings, [210] and any 'Non-Interest Bearing Current Liabilities' (NIBCLs) from capital, [211] to avoid double-counting financing costs. Common financing charges consist of interest expense for debt, preferred dividends, common dividends, and minority interest. NIBCLs include, e.g., accounts payable, dividends payable, accrued taxes, and other accrued liabilities.

Finally, the EVA concept involves several US-GAAP related adjustments, also referred to as "equity equivalents" that *"[...] eliminate accounting distortions by converting from accrual to cash accounting, from a pessimistic lenders' to a realistic shareholder' perspective, and from successful-efforts to full-cost accounting"*:[212]

In view of leasing as equivalent financing method to a purchase with debt and capitalization of debt-financed operating assets, operating leases shall be capitalized.[213] Consequently, capitalized operating leases represent a debt equivalent.

Otherwise distorting the cash performance of the firm, several non-cash reserves are to be reversed. [214] Foremost, the Last-in-First-out (LIFO) inventory reserve reflects historical book value rather than market value that includes inflationary effects on the replacement cost of inventories.[215] Furthermore, several precautionary reserves such as reserves for inventory obsolescence, allowances for doubtful accounts, reserves for warranties, expected sales returns and other similar items, allowances on deferred tax assets, and reserves for future restructurings and closures represent to some extent a management device for earnings management rather than reflecting the timing of charges when actually incurred and, under the EVA concept, are therefore to be reversed.

Assuming that goodwill does not deteriorate, any goodwill shall be fully recognized, which has three implications: Recorded goodwill has to be unamortized, to avoid that

---

[210] See Stewart (1991, p. 86).

[211] See Stewart (1991, pp. 92-93). NIBCLs consist of accounts payable and current accrued expenses and represent capital supply out of the course of business. Implied financing costs of paying suppliers, employees and others yet constitute a portion of cost of goods sold.

[212] See Stewart (1991, p. 91); I do not discuss the 'Equity Equivalent' adjustment converting successful-efforts accounting to full-cost accounting, since this adjustment primarily applies to natural resource firms, e.g. oil drilling companies, that are not the focus of our study.

[213] According to SFAS No. 13, operating leases are treated as rental of property by the lessee. For instructions on the capitalization of operating leases, see Stewart (1991, pp. 98-99).

[214] See Stewart (1991, pp. 113, 117).

[215] The "Last-in, first-out" (LIFO) reserve equals the difference between LIFO and "first-in, first-out" FIFO determined costs of inventories. Assuming an inflationary market, the FIFO method, prevalently used, results in an understated value of inventories captured by the LIFO reserve.

non-cash goodwill amortization imposes bias on the cash-on-cash yield.[216] Additionally, impairment charges are to be reversed, representing one-time write-downs that do not provide information on recurring operating profits.[217] Finally, unrecorded goodwill shall consistently be capitalized unless in the case of a merger of equals.[218]

Moreover, certain expenses, not being expected to generate benefits in the period in which they are recognized, are to be recognized as intangibles, to avoid that long-term investments result immediately in negative earnings signals, while providing benefits to the long-term performance of the firm.[219] Such expenses may include charges for research and development, marketing, training and development of personnel, signing bonuses, customer relationship management systems, supply chain management and knowledge management.

Non-current deferred taxes are a non-interest-bearing income tax liability like a free loan from the government. Assuming that deferrals are infinitely increasing in line with the operations, the EVA concept requires that deferred taxes are treated as an equity equivalent instead of debt, since they represent cash indefinitely invested in the firm that requires earning a return.[220]

Finally, EVA requires recognizing full compensation costs. First of all, this requires recognizing the full obligations and compensation costs of pension and postretirement plans are, to avoid understatement of pension and postretirement expenses excluding investment returns and accruals, and non-capitalization of the complete net pension and postretirement liability that the employing firm has initiated and agreed to serve.[221] Moreover, if the firm carries an 'Employee Stock Ownership Plan' (ESOP), capital costs on ESOP loans are deducted from earnings, to avoid understatement of operating

---

[216] See Stewart (1991, p. 117); notably, there has been some changes in goodwill accounting: Formerly, goodwill has been recorded under the purchase method and annually amortized, following APB 17, paras. 29-30, or not capitalized under the pooling method, see APB 16, par. 45. Nowadays, SFAS No. 141 requires goodwill to be recorded and to undergo annual impairment tests, according to SFAS No. 142.

[217] See Stewart (1991, p. 78).

[218] See Stewart (1991, pp. 114-115); as previously mentioned, today, goodwill is always recorded following SFAS No. 141.

[219] See Stewart (1991, pp. 115-116, 185, 190, 742, 2003, p. 75).

[220] Deferred income taxes arise out of temporary timing differences as cumulative difference between accounting provisions for income taxes and taxes actually paid, see FAS 109. It is likely that firms carry an infinitely increasing deferred tax reserve, since several accounting rules permit higher deductions on tax records than on financial records, e.g. the accelerated depreciation on tax records vs. reported straight-line book depreciation of assets. Tax reserves not expected to be reversed, since assets are constantly replenished according to the going-concern principle, distorts cash performance. For balance sheet classification following the economic model, see Stewart (1991, p. 34).

[221] See Stewart (2003, pp. 69-72). Following US-GAAP, employers record an accrued liability or prepaid expense and recognize pension expense including service and interest costs less the return on plan assets and amortization of several unrecognized items under defined benefit pension plans, see SFAS No. 87. Accounting for defined benefit postretirement plans is largely equivalent, see SFAS No. 106.

compensation expenses.[222] Last, included stock option expenses shall reflect the fair market value of the option on the date of grant rather than a zero intrinsic value, to exclude any gains and losses from exercise or expiration and, thus, to reflect the firm's operating activities rather than the firm's activities as equity holder.[223]

To summarize, Figure 12 outlines NOPAT and capital employed ('Capital') calculation including some key accounting adjustments:

From a financing perspective, NOPAT calculation follows a bottom-up approach, starting with net earnings available to common shareholders, and Capital equals the sum of equity and financial capital components. From an operating view, NOPAT is computed top-down, starting with gross sales, and Capital equals the sum of relevant assets and net working capital. On the basis of the correct application of accounting principles, resulting NOPAT and Capital assumes the same value following any of the two approaches.[224]

Overall, EVA combines the results from management's operating decisions, which affect mostly NOPAT, investment decisions, which influence mainly Capital, and financing decisions, which affect primarily the WACC, into one figure. Consequently, it is a summary measure that *"integrates operating efficiency and balance sheet management in one measure accessible to operating people "*.

---

[222] See Stewart (1991, p. 534). Following US-GAAP, a leveraged ESOP represents a stock bonus plan investing in the employer's stock, debt and accrued interest payable and reporting a funding loan as ESOP liability, see SOP 93-6.

[223] See Stewart (2003, p. 81). Following APB No. 25, stock-based compensation represents another type of a recurring, operating compensation expense. Whereby, ARB 43, Ch. 13B, par. 3, allows recording stock-based compensation expense at the intrinsic value based on the quoted price of the stock at the end of the fiscal year. Thus, issuing stock options to employees generally resulted in recognition of no compensation cost, since options granted at-the-money had an intrinsic value of zero. However, the fair value, calculated via option-pricing models like Black-Scholes, is higher than the intrinsic value due to the time value of the option. In this respect, SFAS No. 148 requires the disclosure of stock-based compensation expense that would have been reported as fair value of the stock options at the grant date. Notably, the adjustment does not apply starting at fiscal year 2006, with the introduction of SFAS 123R requiring stock-based compensation to be measured based on the grant-date fair value of the awards.

[224] See Stewart (1994, p. 73).

Financing Perspective

Income available to common	Common equity
+ Increase in equity equivalents	+ Equity equivalents
[Increase in deferred tax reserve	[Deferred tax reserve
+ Increase in LIFO reserve	+ LIFO reserve
+ Goodwill amortization	+ Cumulative goodwill amortization
+ Increase in net capitalized intangibles	+ Net capitalized intangibles
+ Increase in full-cost reserve	+ Unrecorded goodwill
+ Unusual loss (gain), after-tax	+ Full-cost reserve
+ Increase in other reserves]	+ Cumulative unusual loss (gain), after-tax
+ Preferred dividend	+ Other reserves]
+ Minority interest provision	+ Preferred stock
+ Interest expense after taxes	+ Minority interest
= *NOPAT*	+ All debt
	= *Capital*

Operating Perspective

Sales	Net working capital
- Recurring cash economic operating expenses	[Current assets
- Depreciation	- Non-interest-bearing current liabilities
- Cash operating taxes	- Marketables securities
[Provision for income taxes	- Construction in progress]
- Increase in deferred income tax reserve	+ Net fixed assets
+ Tax savings excluding unusual losses (gains)	[Net property, plant and equipment
+ Tax shield from interest expense on debt	+ Goodwill
+ Taxes on passive investment income]	+ Other long-term operating capital]
= *NOPAT*	+ Equity equivalents
	= *Capital*

Figure 12   EVA Calculations from Financing and Operating Perspectives

2.3.2 Accuracy and Complexity of EVA Definition in Practice

EVA deviates considerably from accounting-based basic RI, due to the wide range of EVA accounting adjustments.[225] All these adjustments towards the economic model reduce distortions of bookkeeping entries. [226] Therefore, EVA should be fully adjusted for external investment analysis, to get the best possible estimate on the firm's long-term abilities to create shareholder value.

However, corporate executives, being resistant to diverge from audited GAAP-based numbers, tend to apply few adjustments, to keep administration costs and complexity low.[227] In fact, EVA implementation usually compromises accuracy and complexity of the definition of EVA.[228]

Consequently, at the beginning Stewart declared to apply mostly 5 to 10 adjustments for EVA clients, keeping the balance between simplicity and precision and minimizing implementation costs and hurdles. [229] Other value-based consultants recommend to managers using 10 to 12 adjustments, and, increasingly, even only six or fewer adjustments.[230] A more recent survey of 29 Stern Stewart EVA clients reveals that firms apply on average 19 adjustments to accounting data, while the number and choice of accounting adjustments differs considerable by firm.[231]

Another survey of 21 large German firms demonstrates that differences in EVA calculation imply considerable imprecision due to inconsistencies and simplifications, leading to substantial misstatements of additions to shareholder wealth, as follows:[232]

• No more than 32 % of surveyed firms calculate EVA based on the standard definition as residual income after taxes. 55% of firms use pre-tax definitions, overestimating value addition. 27% of firms use return ratios, usually neglecting to consider the spread relevant to cost of capital, despite such spread provides the actual value measure in intensive terms.

---

[225] "A system like EVA produces very different results from the SEC accounting rules, which are based on GAAP" (J. M. Stern, 1994, p. 60).

[226] "The accounting model has its uses, but it can also be badly distorted by bookkeeping entries" (J. M. Stern, 1980, p. 7).

[227] Omitted adjustments may also be integrated as supplemental (non-financial) measures in operating plans, strategic plans, and/or compensation programs.

[228] „At some point a trade-off exists between achieving a more accurate return and additional complexity" (G. B. Stewart, 1991, p. 92).

[229] Stern Stewart & Company address about 20 to 25 issues in detail while finally adjusting for only 5 to 10 issues (see G. B. Stewart, 1994, p. 74); a list of adjustments for clients of Stern Stewart & Co., representing the core of their consulting activities, is not publicly available.

[230] See Young (1999).

[231] The survey (Weaver, 2001) lists the following used adjustments: Discontinued operations, extraordinary items, changes in accounting principles, restructuring charges, other one-time charges, interest expense, interest income, cash taxes, goodwill, intangibles, LIFO reserve, R&D expenses, advertising expenses, operating leases, NIBCLs, deferred taxes, and pension liabilities. Notably, wide differences in EVA calculation persist when limiting the sample to one industry: Three food processing firms apply 16, 24, or 31 considerably differing adjustments, respectively.

[232] See Aders (2001).

- 55% of surveyed firms use pre-tax EVA, while 98% of firms use after-tax cost of capital. Consequently, a majority of firms apply an EVA measure that is inconsistently defined.

- 68% of firms apply book values to derive the cost of capital, although the EVA concept suggests using market values. Thus, e.g. a firm underestimates the cost of capital if the market value of equity is higher than the book value of equity, and vice versa.[233]

- 32% of firms apply corporate-wide betas, neglecting specific characteristics of business units or projects.

- 27% of firms using a cost of debt derived from the income statement may not reflect correctly their current market costs.

Equally, the consultancy Stern Stewart & Co. only applies few unspecified standard adjustments when compiling EVA ranking data.[234] Trying to identify these standard adjustments, one study gained the best correlation results of about 90% between independently computed and published EVA data based on 9 adjustments.[235] Another study finds that two adjustments, namely research and development (R&D) expenses and Last-in-First-out (LIFO) reserves, explain on average 92% of the adjustment impact included in Stern Stewart ranking data.[236] Notably, EVA ranking data are not adjusted for depreciation, although literature classifies the depreciation adjustment as one of the major adjustments.[237]

---

[233] Differences in book and market value of equity may be due to the appropriation of earnings, accounting policy, losses carried forward, transfer of profits, etc.

[234] Stewart (1994, p. 73) states: "*In our published rankings and illustrations we have chosen to make only a handful of such adjustments in the calculation of EVA and MVA – typically those which can be made with information contained in the Compustat® database and easily explained to the general business reader. By doing so, we may have inadvertently given the impression that EVA is derived in a mechanical fashion by applying a stereotypical and limited range of adjustments.*"

[235] Yook (1999) applies adjustments for unusual losses and gains, NIBCLs, non-operating assets in terms of marketable securities and CIP, non-capitalized leases, bad debt reserves, LIFO reserves, R&D expenses, and goodwill. Then, he finds correlations from independently computed and published capital, NOPAT, and WACC of 0.92, 0.93, and 0.85, respectively. Low correlations of WACC data most likely refers to differing estimates for beta, the cost of debt, or marginal tax rates.

[236] See Anderson et al. (2004); application of 5 adjustments, namely R&D, advertising, LIFO, bad debt, and operating leases, explain on average 93 percent, whereby the impact of adjustments varies by year and is insignificant.

[237] Major EVA accounting adjustments refer two R&D, deferred taxes, provisions for warranties and bad debt, LIFO reserves, depreciation, goodwill, non-recurring gains and losses such as restructuring charges, and accounting for the capital charge, see O'Byrne and Young (2001, pp. 205-268) and Young (1999).

2.3.3   EVA Applying across Sectors, Business Units, and Management Functions

EVA generally applies across sectors. E.g., EVA had a considerable impact on the performance of the U.S. bank Wells Fargo and the U.S. semiconductor producer Intel.[238] Further, there is evidence on the applicability of EVA in the computers, oil & gas, banking, automotive, and airlines industries as well as in the new economy.[239]

EVA not only summarizes annual financial data from a shareholders' perspective, but also provides *"the basis for the entire range of financial management functions, from capital budgeting, the setting of corporate goals, to shareholder communication and management incentive compensation."* [240] Whereas, an EVA financial management system essentially ties EVA-based performance management to EVA-based compensation plans. [241] Moreover, investors may use EVA as investment analysis instrument.

In the following, key uses of EVA are discussed in more detail to illustrate the dimensions of the concept, namely: performance management, incentive plans, capital structure decisions, and investment analysis.

Performance Management

EVA-based performance management involves recognizing the key principles on value creation in any operating decision. Following the capital spread definition of EVA, namely $EVA_t = (ROCE_t - WACC_t) \cdot Capital_{t-1}$, [242] every firm has five levers to increase shareholder value creation:[243]

- To improve returns from assets yet in business by increasing business revenue or reducing operating expenses.

- To reduce the average cost of capital.

- To invest additional capital in new investments earning above the cost of capital.

---

[238] See Knight (1998, pp. 130-143).

[239] E.g., see Stewart commenting on the value-based performance measure EVA: *"EVA is a generic measure of performance and value. It can apply across businesses, at stages of growth, and whether a business is capital-intensive or a service business."* Walbert (1995, p. 106). Thus, v. Milano (2000) and Antill and Arnott (2004) argue that the EVA method is equally applicable for oldline and emerging new economy firms. Further, several studies propose and test the value relevance of the EVA measure within various specific industries: Milunovich and Tsuei (1996) suggest EVA, providing data for 11 computer software and services firms; McCormack and Vytheeswaran, (1998) recommend EVA for the oil sector, reviewing the largest 25 oil and gas firms; Uyemura, Kantor, and Pettit (1996) suggested EVA for banks, examining the largest 100 bank holding firms; Pettit et al. (2000) advocate EVA for car manufacturers, providing data on 11 automotive firms; finally, Pettit and Goldberg (2000) promote using EVA for airlines, analyzing 8 air carriers. Empirical results are not further discussed, since derived from rather small samples.

[240] See Stern (1994, p. 46); equivalently, see Chew et al. (1996, p. 225).

[241] See Walbert (1995, p. 107);

[242] See section 2.2.2.

[243] See Grant (2003, pp. 77-78), Ross (1995, p. 107), and Stewart (1991, pp. 136-140).

- To increase capital efficiency, optimizing "stay in business" capital expenditures (CAPEX) for extraordinary maintenance and reducing working capital by shortening stock or receivables cycle times.
- To release capital divesting unprofitable assets while preserving core productions and key market positions.

However, these five EVA levers, representing generic value drivers equally applicable to (almost) all business units, lack operational specificity. Therefore, these levers have to be deployed into business-unit specific value drivers and, in the following, decomposed into operating value drivers, tying to specific decisions of frontline management, to reach throughout organizational levels.[244]

Illustratively, the following figure shows a sample EVA value driver tree that contains identified core processes and critical value drivers for an order fulfillment process of a computer manufacturer: [245]

In this example, the company concluded that the value-creating proposition of focusing on identified successful products (computer solutions) has to translate at business-unit level, into market share growth and satisfactory EVA results; however, business unit managers relate share growth and EVA to its elements, namely customer satisfaction, flexibility, and productivity, representing relatively broad measures; department managers within a business unit apply specific work flow measures related to quality, delivery, cycle times, and waste corresponding to their discrete responsibilities and day-to-day decision making.

---

[244] See Copeland et al. (1994, pp. 103-109); whereby, the authors exemplary list generic value drivers, consisting of the EVA margin, composed of a revenue and cost component, and *Capital*, i.e. the working capital, fixed assets, and equity equivalents components, business-unit specific value drivers such as customer mix, sales force productivity, fixed cost (allocations), capacity management, and operational yield, and operating value drivers, such as percent accounts revolving, cost per visit, unit revenues, billable hours to total payroll hours, percent capacity utilized, cost per delivery, accounts receivable terms and timing, and accounts payable terms and timing.

[245] See Lynch and Cross (1995, p. 47).

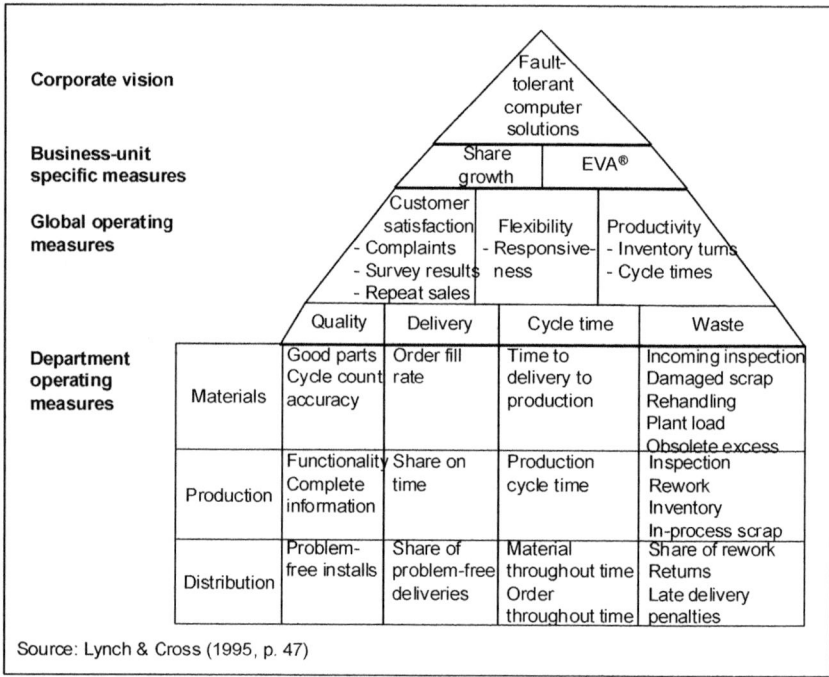

Figure 13   Illustrative EVA Value Driver Tree

Moreover, it is also notable that the EVA concept is applied together with another high level operational target (market share growth in core products) in order to address management actions consistent with the corporate vision for pursuing the value creation paradigm. This is particularly necessary when implementing growth strategies with adoption of a value management system based on EVA that, in particular due to age bias (see section 3.1), implies an intrinsic focus of management on short-term capital efficiency and returns vis-à-vis growth and strategic investments.

So, while financial value drivers like EVA are actually lagging indicators of value creation, forward-looking indicators (non-financial value drivers) have to be associated to the top management view and strategy aimed at maximizing future EVAs, recognizing the continuum from leading to lagging performance indicators:[246] E.g., non-financial learning and growth variables (such as employee skill, morale and suggestions) lead non-financial business process variables (such as product cycle time to market and process quality) that, in turn, lead customer measures (such as

---

[246] See Kaplan and Norton (1996); contrarily, a purely financial value driver system may decompose EVA equivalent to the DuPont analysis.

satisfaction, product quality perceived and on-time delivery), being non-financial value drivers deemed to be the basis for adequate margins and sustainable value creation, traced and measured in lagging EVA results.

Incentive Plans

An EVA-based compensation plan, representing the second core component of an EVA financial management system, aims at aligning management behavior with the interests of shareholders.

Originally, EVA compensation plans were suggested to incorporate the change and level of periodic EVA, as follows: [247]

$$Bonus_t = Multiplier_1 \cdot (EVA_t - EVA_{t-1}) + Multiplier_2 \cdot EVA_t.$$

The EVA improvement is multiplied with a constant multiplier to penalize management with a negative bonus component for decreases in EVA. The current level of EVA is reflected with a multiplier that becomes zero for negative EVAs to reward only additions to shareholder wealth.

More recently, modern EVA bonus plans propose paying a target bonus (that approximates average levels of peers) and a fixed share of EVA improvement in excess of expectations (the expected EVA improvement, EI, equals an EVA level that provides shareholders expected returns), as follows:[248]

$$Bonus_t = Target\ bonus_t + \left(\frac{Target\ bonus_t}{EVA\ interval_t} \right) \cdot [(EVA_t - EVA_{t-1}) - EI_t] =$$

$$= Target\ bonus_t \cdot \left[1 + \frac{(EVA_t - EVA_{t-1}) - EI_t}{EVA\ interval_t} \right],$$

Distinctively, this bonus plan applies an EVA interval that is defined as decline in EVA that delivers a zero shareholder return. Then, the resulting bonus payout pattern matches changes in shareholder wealth, providing unlimited upside potential to management and penalizing for underperformance.[249]

---

[247] *"The essence of EVA bonuses is to pay management a percentage of both the total EVA and the hange in EVA"* (J. M. Stern, 1990, p. 53); *"For companies that aim to increase their competitiveness by decentralizing, EVA is likely to be the most sensible basis for evaluating and rewarding the periodic performance of empowered line people, especially those entrusted with major capital spending decisions"* (Chew et al., 1995); additionally, see Stewart (1991, p. 247).

[248] See Young and O'Byrne (2001, pp. 305-380).

[249] Before adoption, the EVA bonus plan requires a calibration to avoid too much retention risk from multiyear zero bonuses.

Capital Structure Decisions

From a financing view, capital structure decisions may affect EVA via the WACC element. Assuming that the expected return of a levered firm's stock, r_e, follows a linear function of the debt-to-equity ratio, D/E, the cost of capital, $WACC$, is invariant to changes in the capital mix, as follows (with w = respective capital weight of debt or equity capital and r = respective expected return of a firm's debt or equity):[250]

$$WACC_{Levered} = w_d \cdot r_d + w_e \cdot r_e = w_d \cdot r_d + w_e \cdot \left[WACC_{Unlevered} + \left(WACC_{Unlevered} - r_d\right) \cdot \frac{D}{E}\right] =$$

$$= WACC_{Unlevered}$$

Consequently, EVA is not affected by the capital structure. However, this reasoning requires assuming a perfect capital market and no corporate tax deductibility of debt interest expenses. With capital market imperfections, EVA of a levered firm's is higher than EVA of an equivalent business-risk unlevered firm, sine the implicit positive tax subsidy of debt (based on a given tax rate of shareholders, t) lowers the cost of capital:

$$WACC_{Levered} = WACC_{Unlevered} \cdot \left(1 + t_e \cdot \frac{D}{E}\right).$$

Therefore, in practice, changes in the firm's capital structure have an effect on the shareholder value added indicated by EVA.[251]

Investment Analysis

From an investment perspective, EVA stands out as alternative to traditional value or growth styles of investing and market analysis.

Commonly, EVA-based stock analysis compares EVA with 'Market Value Added' (MVA). MVA is defined as the firm's market value of its liabilities and equity less the economic book value of the invested capital, or, equally, the present value of all future EVAs (see Figure 14).[252] The Stern Stewart league table uses ex post MVA calculation to identify top creators of shareholder wealth, see section 2.2.5.[253] Ex ante MVA calculation serves to evaluate stocks.

---

[250] See Modigliani and Miller (1958).
[251] However, tax subsidies may be overestimated, since the debt tax subsidy rate approaches zero under progressive income taxation, see Miller (1977).
[252] *"In theory, a company's market value added at any point in time is equal to the discounted present value of all the EVA, or residual income, it is expected to generate in the future"* (Stewart, 1991, p. 192). *"The difference between a firm's market value and its capital employed, MVA is a measure of the value a company has created in excess of the resources already committed to the enterprise"* (Stewart, 1991, p. 741). While dividend payments do not affect MVA (market and book value of common equity fall simultaneously), MVA is distorted for firms that either spin off or give dividend shares in subsidiary units to their shareholders, see Stewart (1991, p. 196). Noteworthy, MVA is referred to as 'Market Value Lost' if Capital is greater than market value, or, equally, the present value of future EVAs is negative, see Stewart (1991, p. 154). While Stewart (1991) coined the term MVA, it implies that the book value of the firm's assets is adjusted for a number of accounting practices that tend to underestimate invested capital.
[253] *"Shareholders' wealth is maximized only by maximizing the difference between the firm's total value and the total capital investors have committed to it"* (Stewart, 1991, p. 72).

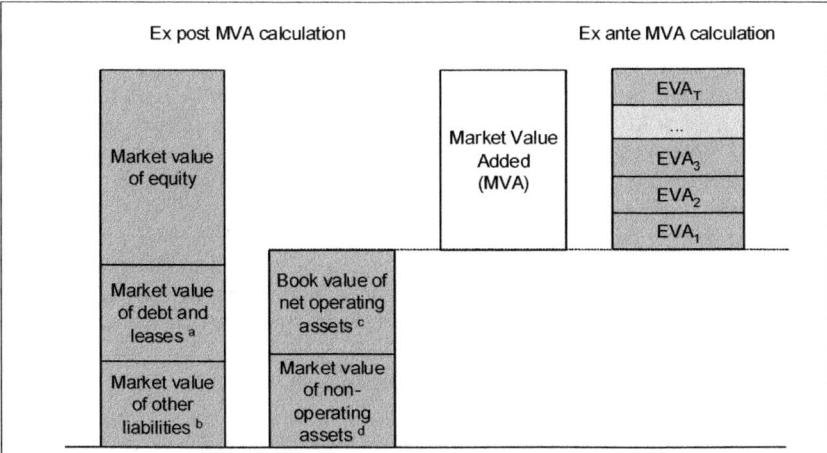

Figure 14   Market Value Added (MVA) Calculation

Stock analysis involves comparing the actual market value and ex ante firm value. The ex ante firm value equals the sum of the present value of future EVAs, the reported net operating assets and the market value of non-operating assets. [254] Differences between ex ante value derived from future EVAs and market value imply that either the equity is undervalued or overvalued or EVA estimates do not match performance expectations of the market.

Instead of relating EVA to market value, a comparison of an EVA-based profitability index to scaled market value alternatively serves to analyze stocks:[255]

Empirical analysis reveals a positive relationship between the market valuation multiple (ratio between market and book value of the capital invested) and the profitability ratio (defined as ROCE divided by the cost of capital) of a company, which may be approximated via a linear, log-linear, or Markowitz model. Consequently, attractive buy opportunities have low corporate valuations in the presence of positive EVA prospects. In comparison with the predicted values from a least-squares fitted relationship, these show a relatively high profitability ratio (higher than one) and a relatively low value-to-capital ratio.

---

[254] See Stewart (1991, p. 153) and Copeland et al. (1994, p. 146).
[255] See Grant (2003).

EVA-based selection of single securities or portfolios can be extended to find potentially attractive industries. Industry analysis aims at finding sectors with maximum EVA prospects for any given level of industry valuation, assuming a positive (linear) association between EVA and MVA, both scaled by capital. Consequently, a MVA-to-Capital ratio lower (higher) than predicted values from a given EVA-to-Capital ratio implies that investors either underestimated (overestimated) the long-term ability of the sector to generate shareholder wealth or, if capital market are largely efficient, that investors were pessimistic (optimistic) about EVA prospects of the industry.

Finally, analysis of the residual return on capital, defined as the economy-wide excess return after the cost of capital, provides information on shareholder wealth creation at the macroeconomic level that explains the country's standard of living.

2.3.4 EVA Perceived as Panacea While Improvements Are Still Needed

There are diverse rationales for corporations to adopt an EVA-based management system. Primarily, EVA is esteemed as metric that creates awareness for the cost of equity, realigns corporate functions, and induces a value-based corporate culture.

Foremost, EVA is commonly adopted to instill the consciousness that equity capital is not for free.[256] In fact, EVA is often misunderstood as fundamental unadjusted residual income, as stated by management theorist Peter F. Drucker: *"EVA is based on something we have known for a long time: what we call profits, the money left to service equity, is not profit at all. Until a business returns a profit that is greater than its cost of capital, it operates at a loss. Never mind that it pays taxes as if it had a genuine profit. The enterprise still returns less to the economy than it devours in resources... Until then it does not create wealth; it destroys it."*[257] In fact, However, EVA differs considerably from basic residual income: It aims to reflect economic rather than accounting value added, excluding unusual and non-operating items and adjusts for a series of accounting principles.[258]

---

[256] See several corporate managers commenting on EVA, e.g., Coca-Cola's CEO Roberto Goizuetta *"You only get richer if you invest money at a higher return than the cost of that money to you. Everybody knows that – but many seem to forget."*(Ross, 1997, p. 116), Allwaste Environmental Services' chief financial officer Wayne Wren *"Of course, at corporate we realized capital had a cost but the concept that equity had a cost was a difficult one for operating people"* (Ross, 1997, p. 117), Herman Miller's chief financial officer Brian Walker *"We really thought capital was free and so the business was having a heart attack."*, or Deere's chief financial officer Pierre Leroy [commenting on the introduction of an EVA-based incentive system] *"People are much focused on the cost of capital rather than the level of sales."* (Ross, 1994, p. 109), and, equivalently, AT&T's chief financial officer Jim Meenan *"Every decision is now based on EVA. The motivation of our business units is no longer just to make a profit. The drive is to earn the cost of capital."* (Walbert, 1994, p. 112).

[257] See Drucker (1995).

[258] See sections 2.3.1 and 2.3.2.

Moreover, the EVA metric stands out as centerpiece of a value-based management system that aligns various corporate functions with shareholder objectives. [259] Commonly, corporations adopt EVA to integrate diverse business decisions via one performance measure. [260] Once implemented across functions and levels, EVA is supposed to introduce a behavioral change, motivating employees to act like shareholders. Indeed, firms report enthusiastically that EVA enabled them to establish a value-based mindset among employees.[261]

While providing a series of benefits, the EVA concept also implies some weaknesses, as follows:

EVA does not adjust for business cycle variations. Therefore, firms in cyclical industries, such as raw materials, cars, chemicals, construction, paper, steel, and heavy equipment, show considerable volatile EVAs.

Furthermore, EVA is less appropriate for industries with (substantially) low capital intensity. Especially for commercial companies providing services mainly based on 'human capital' (rather then on fixed assets) or retailers with highly negative working capital offsetting most of the book value of fixed assets, the Capital Asset Pricing Model (from which is normally derived the cost of equity incorporated in the cost of capital) generally does not capture the underlying commercial risks and actual required equity returns. Then, EVA levels from firms of very low capital intensity sectors such as 'fast moving consumer goods', commercial trading, advisory and engineering normally differ from EVA levels generated in capital-intense sectors such as automobiles, chemicals and oil refinery.

Just as any other VBM measure recognizing the total cost of capital, EVA implies potential errors in estimating the equity cost of capital. Especially, employed historical beta estimates disregard future capital structure and risks and do not adjust for reduced risk if additionally raised equity is used to repay debt.

Furthermore, EVA does not specifically incorporate intangibles such as brand equity and human resources, although it leaves the impression of adjusting for any possible accounting distortion.

Finally, the EVA metric does not treat non-cash charges consistently: NOPAT shall equal *"profits available to provide a cash return to all financial providers of capital to the firm"*, therefore, excluding non-cash taxes and several non-cash charges, e.g.,

---

[259] See sections 0 and 2.3.3.

[260] E.g., Scott Paper's chief financial officer Basil Anderson confirmed that the introduction of EVA resulted due to the limited use of DCF: *"We used to have different financial measures for different purposes – discounted cash flow for capital decisions, another measure for rewarding performance and the like. There was no direct connection between how I evaluated current performance and how I evaluated a project decision. Now EVA is one measure that integrates all that."* (Walbert, 1994, p. 111). Equivalently, Furon's chief financial officer Houdeshell *"We use EVA for capital expenditures, acquisitions, operating unit performance, essentially every key decision that we make."* (Walbert, 1995, p. 106)

[261] See Furon's chief financial officer Houdeshell *"EVA makes a manager of a factory think the same way as an investor."* (Walbert, 1995, p. 106) and Furon's CEO Wills *"EVA creates behavior that allows our folks to be entrepreneurs in their own business activity"* (Walbert, 1995, p. 108).

increases in reserves and goodwill amortization .[262] Thus, it involves hypothetical cash-based taxes exclude tax subsidies from debt capital, taxes on non-operating income, and tax effects induced by deferred taxes. Contradictorily, NOPAT does not exclude non-cash depreciation expenses. Depreciation is deducted to cover foreseeable replacement costs, induced by the use of assets, over the asset life.[263] Equally, 'capital' subtracts accumulated depreciation, while adding back other cumulative non-cash charges, such as accumulated goodwill amortization, and including equity equivalents that reflect the *"economic book value of all cash invested in going-concern business activities"*.[264]

However, deduction of (accumulated) depreciation results in a relative increase of EVA over the asset life. Consequently, biased performance schedule motivate corporate managers to avoid value-adding investments in new fixed assets and to prefer leasing new assets, increasing risk and, thus, the cost of capital.

2.4 Summary

VBM emerged in the 1980s due to increasing pressure from capital markets and shareholders. VBM represents a management system that aims at achieving maximization of shareholder value throughout the organization, with the aid of a certain value-based metric and thereto linked value drivers. To successfully maximize shareholder wealth, VBM firms have to ensure that management is committed, key employees well-trained, processes consistently aligned, and business units accountable and focused on performance. Although a more comprehensive approach is foreseen by other management theories, such as the stakeholder theory and the corporate sustainability concept, shareholder value maximization remains valid as fundamental purpose of the firm.

VBM adoption requires installing a value-based performance metric, since traditional earnings measures are unsuitable to capture shareholder value creation. Earnings (and earnings-based return measures) can be someway manipulated by management, involve non-cash charges, inconsistencies, and volatility, and, above all, do not reconcile to the market value of the firm. In contrast, DCF analysis and thereof derived VBM measures, such as SVA, TBR, RI, EVA, EP, and CVA, follow valuation theory, and, therefore, describe creation of shareholder wealth more accurately. From a theoretical point of view, RI and its variant EVA stand out, representing metrics that reconcile with shareholder wealth and are consistently and feasibly defined based on objective data from annual financial statements. In practice, VBM and value-based

---

[262] See Stewart (1991, p. 86).
[263] *"Depreciation is subtracted because it is a true economic expense. The assets consumed in the business must be replenished before investors achieve a return on their investment. Another way to see this is to observe that a company, when it leases assets, must pay a rent that covers the depreciation the lessor suffers on the lessee's behalf (plus interest)."* (G. B. Stewart, 1991, p. 86).
[264] See Stewart (1991, p. 744).

metrics (especially EVA) are fairly well known and used. VBM metrics can internally be employed for corporate management and communication as well as for the management of business units, divisions and subsidiaries, and externally used for value-based equity investments and performance rankings also used by investment analysts. Empirical studies underline the effectiveness of value-based performance measures: once internally adopted, a corporation commonly alters decision-making to pursue shareholder interests and effectively increases shareholder wealth.

EVA, defined as NOPAT less charges for invested capital, prevails among VBM measures. Contrary to basic RI, EVA calculation involves up to 164 adjustments to accounting-derived earnings and capital, to better reflect economic value creation. However, in practice, corporations do not precisely follow the definition of EVA and apply only few adjustments. Among others, popularity of EVA results from its universal applicability across sectors for financial management and investment analysis. Frequently, EVA is adopted to instill awareness for the cost of capital, to establish a value-based mindset, or to realign all processes towards shareholder value creation. Despite its popularity and benefits, EVA is not a panacea: The metric is not fully adjusted for industry characteristics, intangibles, and non-cash depreciation and most likely implies measurement errors of the cost of (equity) capital.

3 Development of New Value-Based Metrics

This chapter reviews the effectiveness of basic RI (and equally EVA) and the effects of age bias, especially when applied to companies characterized by a relatively high incidence of new investments as well as with yet highly depreciated assets.

In this respect, a methodological analysis on RI (and equally EVA) reveals that measured performance depends on the selected book depreciation method.[265] This bias introduced by accounting depreciation may affect managerial attitude towards strategic (long term) investments and equity portfolios. Consequently, several metrics that aim to reduce depreciation-induced bias are examined. Finally, constituting a primary theoretical contribution of the present study, two depreciation-adjusted VBM metrics, namely 'Cash Residual Income' (CRI) and 'Cash Economic Value Added' (CEVA), are introduced that present more stable and neutral results with respect to the asset age.

3.1 Distortions of Residual Income Caused by Accounting Depreciation

3.1.1 Definition and Relevance of Age Bias

RI and EVA both include charges for accounting depreciation. RI, defined as net earnings after cost of capital (notationally: $RI_t = NE_t^{BI} - WACC_t \cdot NA_{t-1}$), does not adjust for any accounting distortions and, therefore, also not for book depreciation.[266] Although EVA, defined as operating profits after taxes and costs for adjusted capital (notationally: $EVA_t = NOPAT_t - WACC_t \cdot Capital_{t-1}$), includes adjustments for various non-cash accounting charges (such as non-cash provisions and reserves), it does not adjust for book depreciation.[267] Consequently, book depreciation from (long-term) investments affect RI and EVA twofold: Depreciation expenses reduce net earnings and accumulated depreciation reduces net capital and, therefore, the capital charge.

---

[265] The following analysis reviews book depreciation methods permitted under US-GAAP, since EVA has been developed as a response to US-GAAP.
[266] See section 2.2.2.
[267] See section 2.3.4 and Stewart (1991, p. 86).

US-GAAP requires straight-line or accelerated depreciation methods, i.e. sum-of-the-years' digits, declining balance or units-of-production.[268] Therefore, the annual depreciation expense is generally constant (under straight-line depreciation) or decreasing (under accelerated depreciation) over time. In present value terms, admitted 'front-end loaded' straight-line and accelerated methods principally allocate costs disproportionately over the asset's useful life (the period during which an asset is used to generate revenues), charging the early years with most of the cost. Instead, 'back-end loaded' depreciation methods that allocate more costs to later periods, such as annuity or sinking-fund depreciation, are not admitted.[269]

However, permitted book depreciation methods are responsible for a relative increase of EVA and RI over the life of an asset, referred to as 'age bias' or 'old plant / new plant trap'. Age bias results from the combination of the following:

- The capital charge of RI and EVA is relatively high in the first years and relatively low at the end of the asset life (as the net book value becomes lower and lower), ceteris paribus.

- Net earnings of RI and EVA are constant (under straight-line depreciation) or decrease (under accelerated depreciation methods) over the asset life, ceteris paribus.

Consequently, firm performance measured by RI and EVA is lower (higher) than 'effective' value creation in the early (late) period of an investment.

Given that firms usually replace assets quite regularly, keeping the average age of assets fairly constant to ensure the long-run survival of the firm, age bias may be easily misunderstood to be irrelevant in practice. However, in reality, age bias plays a major role, representing a quite heavy disincentive for investments (especially under remuneration plans based on RI or EVA) and providing distorted information to investors.

From a corporate management perspective, age bias plays a major role if management compensation plans are tied to a net residual income measure such as RI and EVA. These metrics both implicitly reallocate the overall value creation of an investment in the last part of the project life, detrimental to the firm performance in the first years. If the incentive package of management is based on such a VBM metric, it introduces an incentive for management to privilege short-term investments notwithstanding the potential to add substantial shareholder value by long-term investments. In practice, one important issue is also that under RI or EVA metrics new investments provide commonly a negative contribution to the firm value in the first year(s). Therefore, biased net residual income measures provide misleading signals to corporate managers to refrain from new value-adding investments and to retain old assets beyond realizing an attractive return. Especially, managers are inclined to reject new value-adding projects, if they expect to leave the company in the short-term. To

---

[268] See SFAS 93, par. 21, and SFAS 109, par. 288a.
[269] See SFAS 92, par. 37. Most likely, these back-end loaded depreciation methods are not accepted, since they require subjective estimates for the return on the investment.

summarize, the mismatch between investment needs to sustain shareholder value and VBM results based on RI and EVA determines unmeant and paradox management behavior.

From an investment management view, age bias is an issue if investors create portfolios from firms with relatively high RI (or EVA). Due to built-in performance improvements, firms with relatively old assets show value added based on the RI metric that is higher than the 'effective' value creation. Consequently, stock selection favors firms with relatively old assets, avoiding firms that just invested in new assets to build up and develop sustainable creation of shareholder wealth. Also if investment managers compare corresponding profitability measures (such as ROCE) against the cost of capital, selected stock portfolios will be equally affected by age bias: underinvesting companies present higher profitability results (and therefore higher spreads) with respect to comparable firms with a higher investment aptitude (being more committed in long term sustainability and growth).

From the above it is clear that the elimination of age bias is important, since otherwise management's investment behavior and investors' portfolio selection may be badly biased.

3.1.2 Pervasiveness of Age Bias under Varying Cases

In the following, the specific impact of US-GAAP depreciation methods on the RI (and EVA) is described. A short review of the applied US-GAAP depreciation methods is hereby provided:

- Under the straight-line (S/L) method, the original investment cost less the salvage value is expensed in equal increments over the useful life. Periodic depreciation, *Dep*, equals net assets, *NA*, less non-depreciable assets, *NDA*, before cumulative amortization and depreciation, *ADA*, divided by the useful life, *T*; equivalently, straight-line depreciation equals net depreciable assets divided by the remaining life span, T^*, notationally:

$$Dep_t^{S/L} = \frac{NA_{t-1} - NDA_{t-1} + ADA_{t-1}}{T} = \frac{NA_{t-1} - NDA_{t-1}}{T^*_t} .$$

- The depreciation charge following the sum-of-the-years'-digits (SoY) depreciation method equals the gross value of depreciable assets, *GDA*, times the number of residual years of use, T^*, divided by the sum of the natural numbers until the useful life (in years) of the underlying investment, notationally: [270]

$$Dep_t^{SoY} = GDA_{t-1} \cdot \frac{T^*_t}{T \cdot (T+1)/2} .$$

---

[270] Sum-of-the-years'-digits depreciation applies only to tangible assets with a useful life equal to or greater than three years and differs based on the used averaging convention (i.e., midmonth, modified half-year and half-year).

- Declining balance depreciation is determined upon a predefined depreciation factor, such as 150% and 200%; thereupon, depreciation equals the beginning book value of the (depreciable) asset times the depreciation factor divided by the asset's useful life; the method requires switching to straight-line depreciation when deductions allowed by the straight-line method equal or exceed the deductions allowed by the declining balance method.[271] Using 200% as depreciation factor, double declining (DD) balance depreciation initially requires multiplying beginning book capital by two divided by the useful life, notationally:

$$Dep_t^{DD} = 2 \cdot \frac{NA_{t-1} - NDA_{t-1}}{T}.$$

To illustrate age bias, a case example assumes an initial investment of 16 m USD in new equipment and 2 m USD in working capital, a project and asset life of 7 years, annual operating profits of 3,486.5 k USD, and a cost of capital of 8%. Consequently, the internal rate of return of the project equals 10%; thus, returns above the cost of capital imply that shareholder wealth increases over the project life.

As shown in Figure 15, RI defined after US-GAAP depreciation always improves over the asset life, although cash performance shows no additional value creation.

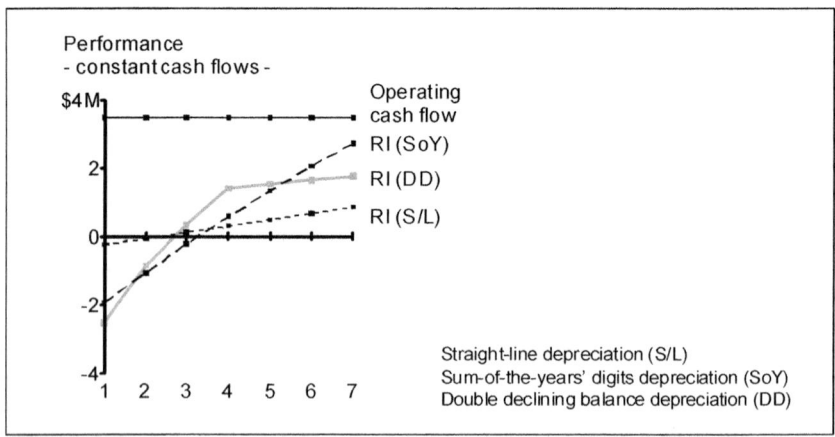

Figure 15   Age Bias of RI – Constant Cash Flows

Table 4 summarizes the underlying RI calculations of the case example, highlighting the performance improvements attributable to age bias under differing depreciation methods.

---

[271] Declining balance depreciation applies only to tangible assets with a useful life equal to or greater than three years. The method is differs, depending on which percentage factor and which averaging convention is used (i.e., midmonth, modified half-year and half-year).

Time		0	1	2	3	4	5	6	7
Constant operating cash flows			3,487	3,487	3,487	3,487	3,487	3,487	3,487
RI (S/L)	Net assets, EOY	18,000	15,714	13,429	11,143	8,857	6,571	4,286	2,000
	S/L depreciation		2,286	2,286	2,286	2,286	2,286	2,286	2,286
	Residual income		-239	-56	127	309	492	675	858
	PI			183	183	183	183	183	183
RI (SoY)	Net assets, EOY	18,000	14,000	10,571	7,714	5,429	3,714	2,571	2,000
	SoY depreciation		4,000	3,429	2,857	2,286	1,714	1,143	571
	Residual income		-1,954	-1,062	-216	584	1,338	2,047	2,709
	PI			891	846	800	754	709	663
RI (DD)	Net assets, EOY	18,000	13,429	10,163	7,831	6,373	4,915	3,458	2,000
	DD depreciation		4,571	3,265	2,332	1,458	1,458	1,458	1,458
	Residual income		-2,525	-853	341	1,402	1,519	1,636	1,752
	PI			1,672	1,194	1,061	117	117	117

Table 4   Performance Improvements of RI – Constant Cash Flows

Under straight-line depreciation, built-in performance improvements of RI are constant over time, namely 183 k USD, due to the progressive increase of accumulated straight-line depreciation, reducing net assets and, thus, capital charges.

Age bias persists (and gets even worse) applying other US-GAAP depreciation methods than straight-line depreciation. Under accelerated methods, performance improvements are considerably higher and decreasing over the asset life, since both depreciation expense and net asset book value (and so, capital charges) decrease over time.

Alternatively, depreciation-based performance improvements, PI, can be calculated based on the depreciation expense, Dep, of the current and previous period and the current weighted average cost of capital, WACC, as follows:

$$PI_t = Dep_{t-1} \cdot (1 + WACC) - Dep_t.$$

In the case of straight-line depreciation, the formula can be abbreviated as: $PI_t = Dep \cdot WACC_t = (GDA/T) \cdot WACC_t$, where GDA = gross depreciable assets and T = the assets' useful live. In this case, the constant annual improvement accounts to 183 k USD, as follows: $PI = [(18,000 - 2,000)/7] \cdot 0.08 = 183$.

To estimate performance improvements under accelerated methods, the basic formula has to be applied. Under sum-of-the-years' digits depreciation, depreciation expenses decreases from 891 m USD in year 1, calculated as $PI_1 = [16,000 \cdot (7/28)] \cdot (1.08) - [16,000 \cdot (6/28)] = 891$, to 663 m USD in year 7, computed as $PI_7 = [16,000 \cdot (2/28)] \cdot (1.08) - [16,000 \cdot (1/28)] = 663$. Under double-declining balance depreciation, the built-in performance improvement decreases annually until year 5, when the method suggests switching to straight-line depreciation. Accordingly, performance improvements decrease from 1,672 k USD, calculated as, $PI_1 = [16,000 \cdot (1/7) \cdot 2] \cdot (1.08) - [11,429 \cdot (1/7) \cdot 2] = 1,672$ to 117 k USD, computed as $PI_7 = [1,458/1] \cdot 0.08 = 117$.

Of course, age bias, as presented above, also prevails under increasing or decreasing cash flows, as shown by the following modifications to the case example above.

To begin with, the case is extended to assume that operating cash flow increases or decreases constantly by 200 k USD per year. Again, net RI annually after book depreciation improves relative to the operating cash flows, as shown in Figure 16.

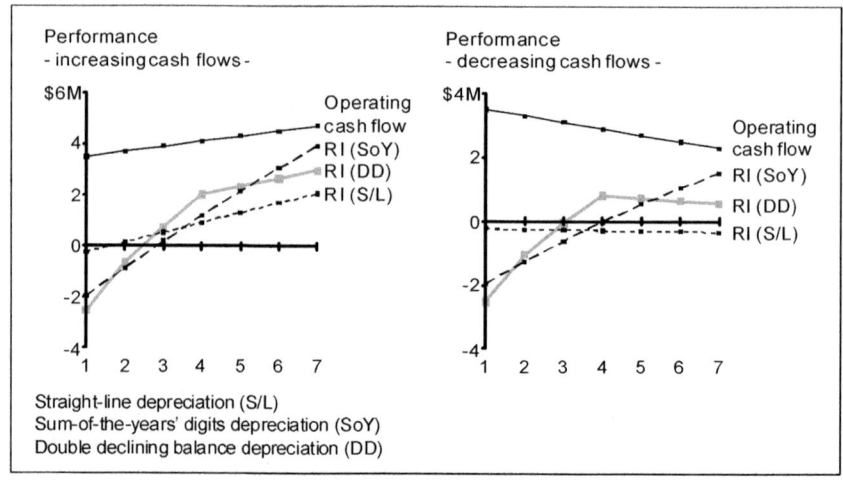

Figure 16   Age Bias of RI – Cash Flow Changes by Fixed Amount

As under constant cash flows, RI net of accelerated depreciation implies larger performance improvements and, therefore, approaches cash performance more rapidly than RI net of straight-line depreciation.

In fact, built-in improvements do not differ under constant, increasing or decreasing cash flows, see Table 5, as indifference of cash flow schedules is yet suggested by the formula, introduced above, that determines performance improvements only based on the cost of capital and depreciation expenses.

Time		0	1	2	3	4	5	6	7
Increasing oper. cash flows			3,487	3,687	3,887	4,087	4,287	4,487	4,687
RI	Net assets, EOY	18,000	15,714	13,429	11,143	8,857	6,571	4,286	2,000
(S/L)	S/L depreciation		2,286	2,286	2,286	2,286	2,286	2,286	2,286
	Residual income		-239	144	527	909	1,292	1,675	2,058
	PI			183	183	183	183	183	183
RI	Net assets, EOY	18,000	14,000	10,571	7,714	5,429	3,714	2,571	2,000
(SoY)	SoY depreciation		4,000	3,429	2,857	2,286	1,714	1,143	571
	Residual income		-1,954	-862	184	1,184	2,138	3,047	3,909
	PI			891	846	800	754	709	663
RI	Net assets, EOY	18,000	13,429	10,163	7,831	6,373	4,915	3,458	2,000
(DD)	DD depreciation		4,571	3,265	2,332	1,458	1,458	1,458	1,458
	Residual income		-2,525	-653	741	2,002	2,319	2,636	2,952
	PI			1,672	1,194	1,061	117	117	117
Decreasing oper. cash flows			3,487	3,287	3,087	2,887	2,687	2,487	2,287
RI	Net assets, EOY	18,000	15,714	13,429	11,143	8,857	6,571	4,286	2,000
(S/L)	S/L depreciation		2,286	2,286	2,286	2,286	2,286	2,286	2,286
	Residual income		-239	-256	-274	-291	-308	-325	-342
	PI			183	183	183	183	183	183
RI	Net assets, EOY	18,000	14,000	10,571	7,714	5,429	3,714	2,571	2,000
(SoY)	SoY depreciation		4,000	3,429	2,857	2,286	1,714	1,143	571
	Residual income		-1,954	-1,262	-616	-16	538	1,047	1,509
	PI			891	846	800	754	709	663
RI	Net assets, EOY	18,000	13,429	10,163	7,831	6,373	4,915	3,458	2,000
(DD)	DD depreciation		4,571	3,265	2,332	1,458	1,458	1,458	1,458
	Residual income		-2,525	-1,053	-59	802	719	636	552
	PI			1,672	1,194	1,061	117	117	117

Table 5   Performance Improvements of RI – Cash Flows Increasing by Fixed Amount

However, under non-constant cash flows, depreciation-induced performance improvements are no longer easy to read from changes in RI. Then, performance improvements equal changes in RI in excess of cash flows changes. E.g., performance improvements under straight-line depreciation remain 183 k USD per year, calculated as change in RI less change in operating cash flow; for instance, performance improvements in year 2 equal:

$$PI_2 = (RI_2 - RI_1) - (\text{Operating Cash Flow}_2 - \text{Operating Cash Flow}_1) =$$

$$= (144 + 239_1) - (3,687 - 3,487) = 183.$$

Equally, age bias prevails if cash flows do increase or decrease by a constant percentage factor rather than by a fixed amount. Therefore, the prior case is extended to include cash flows that increase or decrease by 20% each year. As shown in Figure 17 and Table 6, prior annual performance improvements of residual income after accounting depreciation still prevail.

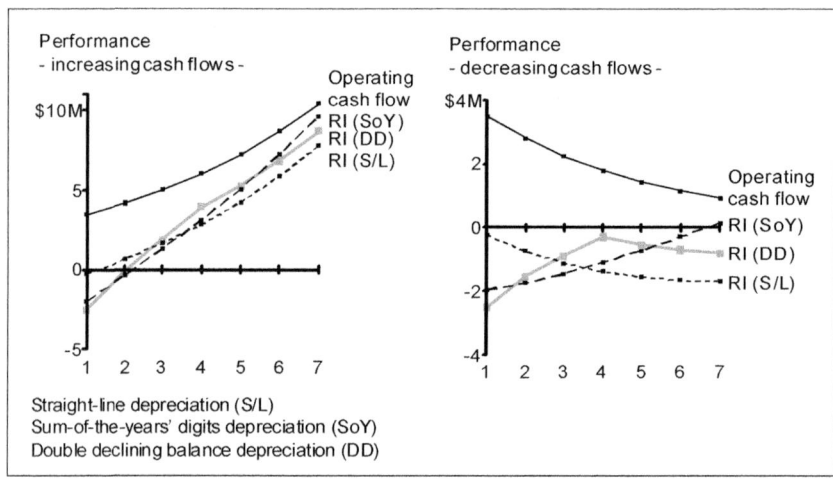

Figure 17   Age Bias of RI – Cash Flow Changes by Fixed Factor

Time		0	1	2	3	4	5	6	7
Increasing oper. cash flows			3,487	4,184	5,021	6,025	7,230	8,676	10,411
RI (S/L)	Net assets, EOY	18,000	15,714	13,429	11,143	8,857	6,571	4,286	2,000
	S/L depreciation		2,286	2,286	2,286	2,286	2,286	2,286	2,286
	Residual income		-239	641	1,661	2,848	4,235	5,864	7,782
	PI			*183*	*183*	*183*	*183*	*183*	*183*
RI (SoY)	Net assets, EOY	18,000	14,000	10,571	7,714	5,429	3,714	2,571	2,000
	SoY depreciation		4,000	3,429	2,857	2,286	1,714	1,143	571
	Residual income		-1,954	-365	1,318	3,122	5,081	7,236	9,633
	PI			*891*	*846*	*800*	*754*	*709*	*663*
RI (DD)	Net assets, EOY	18,000	13,429	10,163	7,831	6,373	4,915	3,458	2,000
	DD depreciation		4,571	3,265	2,332	1,458	1,458	1,458	1,458
	Residual income		-2,525	-156	1,875	3,940	5,262	6,825	8,676
	PI			*1,672*	*1,194*	*1,061*	*117*	*117*	*117*
Decreasing oper. cash flows			3,487	2,789	2,231	1,785	1,428	1,142	914
RI (S/L)	Net assets, EOY	18,000	15,714	13,429	11,143	8,857	6,571	4,286	2,000
	S/L depreciation		2,286	2,286	2,286	2,286	2,286	2,286	2,286
	Residual income		-239	-754	-1,129	-1,392	-1,566	-1,669	-1,715
	PI			*183*	*183*	*183*	*183*	*183*	*183*
RI (SoY)	Net assets, EOY	18,000	14,000	10,571	7,714	5,429	3,714	2,571	2,000
	SoY depreciation		4,000	3,429	2,857	2,286	1,714	1,143	571
	Residual income		-1,954	-1,759	-1,471	-1,118	-721	-298	137
	PI			*891*	*846*	*800*	*754*	*709*	*663*
RI (DD)	Net assets, EOY	18,000	13,429	10,163	7,831	6,373	4,915	3,458	2,000
	DD depreciation		4,571	3,265	2,332	1,458	1,458	1,458	1,458
	Residual income		-2,525	-1,550	-914	-299	-540	-709	-820
	PI			*1,672*	*1,194*	*1,061*	*117*	*117*	*117*

Table 6   Performance Improvements of RI – Cash Flows Increasing by Fixed Factor

To summarize, RI (and likewise EVA) implies built-in annual performance improvements over the asset life and is therefore biased. This so-called age bias prevails under various front-end loaded depreciation methods according to US-GAAP, regardless whether cash performance is constant, increasing or decreasing. Age bias under accelerated depreciation methods is higher than under straight-line depreciation. Due to the persistence of age bias under varying assumptions, some common depreciation-adjusted VBM metrics, introduced in the next section, were developed to reduce age bias.

3.2 Adjustment Techniques Using Figurative Economic Depreciation

3.2.1 Available Metrics Replacing Book Depreciation

Age bias from accounting depreciation suggests removing book depreciation expenses from value metrics. However, depreciation itself, on one side, is essential in providing mathematical reconciliation with corporate value provided under the DCF theory and, on the other side, is a legitimate charge to earnings, since assets wear out or become obsolete requiring replacements to sustain the firm's cash flows.

In this view, a figurative economic depreciation, ED, may be introduced in order to replace book depreciation and account for replacement costs avoiding age bias. Such Economic Depreciation represents the value of a replacement liability.

In the literature, there are a few depreciation-adjusted metrics that replace book depreciation with figurative non-accounting depreciation. Generally, this economic depreciation reflects the asset's actual decrease in economic value over its useful life or, equally, the amount necessary to finance replacement investments. It is assumed to be positive, relating to the deterioration of 'consumable' assets. [272] However, depreciation-adjusted metrics differ on the applied ED calculation methodology.

Reviewing the definition and performance impact of differing economic depreciation methods, four common depreciation-adjusted VBM measures are discussed in chronological order, namely:

- Cash Value Added (CVA)
- 'Revised' Economic Profit (EP)
- Economic Margin (EM)
- 'Revised' Economic Value Added (EVA)

The analysis uses the prior case example to illustrate differences between 'unadjusted' EVA and the depreciation-adjusted measurement concepts. Notably, all five metrics correctly quantify total additions to shareholder value over the project life of 1,319 k USD.

'Unadjusted' Economic Value Added (EVA)

EVA, defined as $EVA_t = NOPAT_t - WACC_t \cdot Capital_{t-1}$, and correspondingly measured profitability, defined as $ROCE_t = NOPAT_t \,/\, Capital_{t-1}$, both include accounting depreciation. [273] Therefore, EVA includes a built-in performance improvement, as shown by increasing EVA s and ROCEs over the investment life (see following table). [274]

---

[272] Negative economic depreciation on appreciating assets, mostly intangibles such as acquisitions and R&D, is neither evaluated in this review nor applied in the succeeding study.

[273] See Stewart (1991; 1994, p. 81).

[274] Fur further details on the EVA metric, see section 5.1.1, and on built-in performance improvements, see section 3.1.

Year	0	1	2	3	4	5	6	7
Capital, EOY	18,000	15,714	13,429	11,143	8,857	6,571	4,286	2,000
S/L depreciation		2,286	2,286	2,286	2,286	2,286	2,286	2,286
NOPAT		1,201	1,201	1,201	1,201	1,201	1,201	1,201
EVA		-239	-56	126	309	492	675	858
ROCE		6.7%	7.6%	8.9%	10.8%	13.6%	18.3%	28.0%

Table 7   Summary of EVA and ROCE Calculation

Increasing measures over time result from the following: Profits after accounting depreciation of 1,201 k USD are constant along the project life, while net capital steadily declines from 18,000 k USD to 2,000 k USD. Applying a cost of capital of 8.0%, EVA increases from -239 k USD in the first year to 858 k USD in the last year due to the decline in capital charges, misleadingly indicating an increase in value addition over the project life. Likewise, the corresponding return ratio ROCE increases from 7% to 28%.

To conclude, measured value added and returns are understated in the initial investment periods (and may be even negative, as in this example) and progressively overstated as the investment is depreciated in value over time.

Cash Value Added (CVA)

The 'Cash Value Added' (CVA) metric includes economic depreciation, *ED*, that equals beginning gross depreciable assets, *GDA*, times a factor determined by the reinvestment rate (i.e., the opportunity cost of capital or real cost of the investment), *r*, and the investment's total useful life, *T*, as follows:[275]

$$ED^{CVA} = GDA \cdot \frac{r}{(1+r)^T - 1}.$$

GDA consist of inflation-adjusted gross plant and equipment, construction in progress (CIP), inflation-adjusted value of non-capitalized leased property and intangibles excluding capitalized pension benefits.[276] Asset life is approximated by dividing depreciation and amortization expense less goodwill amortization by gross plant less land, improvements and CIP.[277] The reinvestment rate is assumed to be constant over time.

Accordingly, economic depreciation, also referred to as sinking-fund depreciation, is constant over the asset life and, under certain circumstances,[278] it represents an annuity that finances replacing depleted investments at the end of the asset's useful life. More-

---

[275] See Boston Consulting Group (1994), Guthrie and Lemon (2004), and Stelter (1999, p. 234).
[276] See Madden (1999, p. 113).
[277] See Madden (1999, pp. 115-124).
[278] Assuming that the reinvestment rate *r* (to calculate economic depreciation) is equal to the annuity discount factor.

over, it is considerably lower than straight-line depreciation and is not deducted from the initial capital invested. In contrast, economic depreciation is compounded over the asset life, accounting for the time value of money. Consequently, annual economic depreciation, compounded at the opportunity cost of capital grows in value over the project life and finally equals the original cost of the investment (i.e., gross depreciables, GDA):

$$GDA = ED^{CVA} \sum_{t=0}^{T-1} (1+r)^t .$$

The overall CVA measure is defined as gross operating cash flows, $OGCF$, less economic depreciation, ED, and capital charges based on inflation-adjusted gross investments, GI: $CVA_t = OGCF_t - ED_t - WACC_t \cdot GI_{t-1}$.[279]Consequently, CVA disregards book depreciation, using economic 'annuity' depreciation to account for deteriorations in the assets' value. Instead, book depreciation is added back to operating profits and cumulative book depreciation is added back to net invested capital. At last, the return measure 'Cash Flow Return on Investments' (CFROI) can be approximated by dividing gross operating cash flows, $OGCF$, after economic depreciation, ED, by gross investments, GI, notationally:

$$CFROI_t = \frac{OGCF_t - ED_t^{CVA}}{GI_{t-1}} .[280]$$

In the following, the case computations (as summarized by the table below) highlight that neither adjusted depreciation expense nor the performance schedule according to the CVA or CFROI measure is affected by the age of the assets.

Year	0	1	2	3	4	5	6	7
Gross investm., EOY	18,000	18,000	18,000	18,000	18,000	18,000	18,000	18,000
Econ. depreciation		1,793	1,793	1,793	1,793	1,793	1,793	1,793
OGCF		3,486	3,486	3,486	3,486	3,486	3,486	3,486
CVA		253	253	253	253	253	253	253
CFROI		9.4%	9.4%	9.4%	9.4%	9.4%	9.4%	9.4%

Table 8   Summary of CVA and CFROI Calculation

Although the CVA method suggests using the five-year past median of the economy CFROI as reinvestment rate [281], in our calculation we shall assume it to be equal to the average cost of capital. Then, annual economic depreciation equals 1,687 k USD:

$$ED^{CVA} = 16,000 \cdot \frac{0.08}{(1+0.08)^7 - 1} = 1,793 .$$

---

[279] See section 2.2.2.
[280] See section 2.2.2.
[281] See Martin and Petty (2000, p. 120).

By definition, economic depreciation of 1,687 k USD per year adds up to the initial investment, as follows:

$$1,793 \sum_{t=0}^{7} (1+0.08)^t = 16,000$$

Periodic shareholder value added by the project equals 253 k USD, as follows:
$CVA = 3,486 - 1,793 - (0.08 \cdot 18,000) = 253$.
The project's periodic profitability provided to shareholders equals 9.4%:
$$CFROI = \frac{3,486 - 1,793}{18,000} = 9.4\%.$$

Consequently, CVA and CFROI are constant over the useful life of the investment, avoiding age bias and, therefore, providing less biased information to management.[282]

'Revised' Economic Profit (EP)

The revised EP metric replaces accounting depreciation with economic depreciation just as CVA, but maintains the net earnings and net capital basis of the basic EP metric.[283]

Economic depreciation, ED, in this case also referred to as present-value depreciation, is defined as the decline in the value of the assets described by the decrease in the present value of the investment's future free cash flows, FCF, discounted at the internal rate of return, IRR, or opportunity cost of capital, as follows:[284]

$$ED_t{}^{EP} = \sum_{j=t}^{n} \frac{FCF_j}{(1+IRR)^j} - \sum_{j=t+1}^{n} \frac{FCF_j}{(1+IRR)^{j-1}}.$$

This equation can be rewritten as: $ED_t{}^{EP} = FCF_t - IRR \cdot NDA_{t-1}$.

Present value depreciation, which doesn't compound over time, differs from sinking-fund depreciation, introduced above, in two ways: First, it sums up to the initial investment cost (i.e., gross depreciable assets, GDA) over the investment horizon, notationally:

$$GDA_0 = \sum_{t=1}^{T} ED_t^{EP}.$$

---

[282] *"The CVA calculation shows a value equal to zero each year, which accurately reflects the underlying reality that the plant is consistently generating a cost of capital return. The adjustments incorporated in CVA avoid the behavioral biases inherent in the EVA calculation in both early and later years and provide the right signals to operating management."* (Boston Consulting Group, 1994, p. 16).

[283] See O'Byrne (2000); also Grant (2003, pp. 207-216) presents this economic depreciation adjustment.

[284] See Martin and Petty (2000, pp. 140-142) and O'Byrne (2000); omitting explicit future cash flow estimates, economic depreciation can be rewritten as expected total return on the investment that equals the expected cash flow less the expected return on the beginning net investments, as follows: $ED_t^{EP} = FCF_t - IRR \cdot NDA_{t-1}$. However, the IRR still implies future cash flow estimates.

Second, economic depreciation relatively increases over the asset life (given constant cash flows).

The overall revised EP measure is defined as net operating profits less adjusted taxes, $NOPLAT$, and accounting depreciation, Dep, less economic depreciation, ED, and the weighted average cost of capital, $WACC$, times beginning net invested capital, IC, notationally $EP_t = NOPLAT_t + Dep_t - ED_t^{EP} - WACC_t \cdot IC_{t-1}$. Thus, accounting depreciation is added back and replaced by economic depreciation. As a measure of profitability, ROIC is defined by dividing depreciation-adjusted net earnings by beginning net capital, as follows:

$$ROIC_t = \frac{NOPLAT_t + Dep_t - ED_t^{EP}}{IC_{t-1}}.$$

Accordingly, revised EP and ROCE are calculated as follows (see Table 9):

Based on the prior case, economic depreciation in the initial year equals 1,686 k USD and increases to 2,988 k USD in the seventh year, computed as follows:

$$ED_t = 3,486 - 0.10 \cdot NDA_{t-1}.$$

Economic profit includes non-constant economic depreciation while adding back accounting depreciation, as follows:

$$EP_t = 1,201 + 2,286 - ED_t^{EP} - 0.08 \cdot IC_{t-1}$$

Overall, increasing economic depreciation expenses dominate decreasing capital charges, resulting in a decrease in EP from 360 k USD in the first year to 100 k USD in the seventh year. In this example, EP is positive, since the project IRR of 10% is higher than the cost of capital of 8%. If the cost of capital equaled the IRR, EP would be zero. If the cost of capital were greater than the IRR, EP would be negative. Under constant cash flows, positive EP decreases and negative EP increases over the asset life, ceteris paribus.

Finally, ROIC constantly equals the IRR of 10%, since enumerator and denominator decrease concurrently over the asset life, as follows:

$$ROIC_t = \frac{1,201 + 2,286 - ED_t^{EP}}{IC_{t-1}} = 10.0\%.$$

Based on economic depreciation, the concept succeeds in indicating constant returns (as the ROIC is constant). However, indicated value additions to shareholders (measured by EP) are biased, declining substantially over the asset's life even when cash flows are supposed to be stable. And most important, economic depreciation accounted in the revised EP metric is biased, since it is determined from subjective expected IRR and free cash flows estimates.

Year	0	1	2	3	4	5	6	7
Investm. capital, EOY	18,000	16,314	14,458	12,418	10,173	7,704	4,988	2,000
Econom. depreciation		1,686	1,855	2,041	2,245	2,469	2,716	2,988
NOPLAT		1,800	1,631	1,446	1,242	1,017	770	499
"Revised" EP		360	326	289	248	203	154	100
"Revised" ROIC		10.0%	10.0%	10.0%	10.0%	10.0%	10.0%	10.0%

Table 9   Summary of 'Revised' EP and ROIC Calculation

Economic Margin (EM)

The 'Economic Margin' (EM) concept claims to combine advantages of the earnings-based EVA and cash-based CFROI concept, while eliminating inherent disadvantages.[285]

Contrary to CVA and revised EP, the EM concept does not include an explicit economic depreciation expense. However, an adjusted economic depreciation expense, referred to as 'return of capital', is automatically included in the 'capital charge', $CapChg$, next to the 'return on capital', RoC, which is defined as beginning gross invested capital, GIC, times the weighted average cost of capital, $WACC$. The capital charge equals the annuity payment based on the weighted average cost of capital, $WACC$, as discount rate, the total asset life, T, as investment life, gross invested capital, GIC, as initial investment, and non-depreciating assets, NDA, as terminal value. Consequently, economic depreciation equals capital charges less return on capital, as follows:

$$ED_t^{EM} = CapChg_t - RoC_t = PMT(WACC_t; T; GIC_{t-1}; NDA) - GIC_{t-1} \cdot WACC_t.$$

'Economic Profit' (EP), the measure of value added, equals operating cash flow, OCF, less the capital charge, notationally: $EP_t = OCF_t - CapChg_t$.[286] The return ratio 'Economic Margin' (EM) is defined as EP divided by gross invested capital, GIC, as follows: $EM_t = EP_t / GIC_{t-1}$.

Based on the case example, adjusted depreciation expense equals 1,793 k USD:

$$ED = PMT(0.08; 7; 18,000; 2,000) - 18,000 \cdot 0.08 = 1,793$$

While economic depreciation is equivalent to economic depreciation included in the CVA concept, it is 'hidden' in the overall capital charge. EP is based on constant gross

---

[285] See Obrycki and Resendes (2000). First of all, EM overcomes EVA's reliance on net assets that introduces problems to compare firms with differing depreciation methods or with differing depreciation life periods. Additionally, EM avoids estimating average useful lives for economic depreciation as required by the CFROI concept. Moreover, EM overcomes communication issues of the internal rate of return measure CFROI, which represents a non-linear measure that does not integrate a hurdle rate. Finally, EM is not affected by leverage as is CFROI that adds back the tax benefit from debt financing to cash flow.

[286] Contrary to the cash flow statement line item 'Cash Flow from Operations', CFO, operating cash flows, OCF, refer to the definitions introduced by the EM concept.

cash flows and constant capital charges, indicating an annual addition of 253 k USD to shareholder wealth, computed as follows:

$EP = 3,486 - 3,233 = 253$

Accordingly, the EM and CVA concept also measure equivalent amounts of value added, denoted as EP or CVA, respectively.

Finally, dividing constant EP by constant gross capital results in a constant EM:

$$EM = \frac{253}{18,000} = 1.4\%.$$

Contrary to the prior profitability measures CFROI and ROIC, EM yet deducts the cost of capital. Therefore its sign indicates additions to or subtractions from shareholder wealth. Approximately, EM equals CFROI less the cost of capital. $EM = 9.4\% - 8.0\% = 1.4\%$.

As shown by Table 10, adjusted depreciation as well as the absolute and relative measures of value creation is constant over time. While EP and EM eliminate built-in improvements of accounting depreciation, surrogate economic depreciation is not considered explicitly.

Year	0	1	2	3	4	5	6	7
GIC, EOY	18,000	18,000	18,000	18,000	18,000	18,000	18,000	18,000
CapChg		3,233	3,233	3,233	3,233	3,233	3,233	3,233
- Return on capital		1,440	1,440	1,440	1,440	1,440	1,440	1,440
- Return of capital		1,793	1,793	1,793	1,793	1,793	1,793	1,793
OCF		3,486	3,486	3,486	3,486	3,486	3,486	3,486
EP		253	253	253	253	253	253	253
EM		1.4%	1.4%	1.4%	1.4%	1.4%	1.4%	1.4%

Table 10   Summary of EP and EM Calculation

'Revised' Economic Value Added (EVA)

While the CVA and EM concept both reduce age bias by converting RI onto a gross earnings and gross asset basis before accounting depreciation, the 'revised' EVA metric maintains a net earnings and net asset basis, just like revised EP.[287]

Economic depreciation is defined as the total cost of ownership, *TCO*, less capital charges (based on the weighted average cost of capital, *WACC*) for beginning net capital, *Capital*:

$$ED_t^{EVA} = TCO_t - WACC_t \cdot Capital_{t-1}.$$

TCO represents a level lease payment that recovers the initial investment and the cost of capital over the asset's life, just as if investments were sold and leased back.

---

[287] See Stewart (2002).

The initial investment is interpreted as principal amount of an annuity loan. Then, TCO is the annuity payment, as follows: $TCO = PMT(WACC_t; T; Capital_{t-1}; NDA)$. The sum of discounted TCOs over the asset life and discounted non-depreciable assets, NDA, equals initial investments, notionally:

$$Capital_0 = \sum_{t=1}^{T} \frac{TCO_t}{(1+WACC)^t} + \frac{NDA}{(1+WACC)^T}.$$

Consequently, TCO is equivalent to the CapChg component of the EM concept.

Economic depreciation increases over the assets' useful life, since TCO is constant and net-based capital charges decrease over the project life due to a decline in net book value. Equivalently, adjusted depreciation can be computed as beginning net depreciable assets times a factor determined by the cost of capital and remaining asset life, T^*, as follows:

$$ED_t^{EVA} = \left(Capital_{t-1} - NDA_{t-1}\right) \cdot \frac{WACC}{(1+WACC)^{T^*} - 1}.$$

Compared to economic depreciation based on gross assets (such as within CVA or EM), economic depreciation based on net assets adds complexity, requiring annual estimates of net capital after cumulative economic depreciation.

Then, revised EVA equals cash operating profits after taxes, $COPAT$, less TCO: $EVA_t = COPAT_t - TCO_t$. Correspondingly, the profitability measure ROCE equals COPAT less economic depreciation divided by beginning capital after cumulative economic depreciation, notionally:

$$ROCE_t = \frac{COPAT_t - ED_t^{EVA}}{Capital_{t-1}}.$$

Given this case, economic depreciation increases from 953 k USD in the initial year to 3,099 k USD in the final year, according to the following formula:
$$ED_t^{EVA} = 3,233 - 0.08 \cdot Capital_{t-1}.$$
Equivalently, adjusted depreciation can be computed as:
$$ED_t^{EVA} = NDA_{t-1} \cdot \frac{0.08}{(1+0.08)^{T^*} - 1}.$$

Whereas, figurative depreciation is part of the constant TCO charge of 3,233 k USD, calculated as: $TCO = PMT(0.08; 7; 18,000; 2,000) = 3,233$. By definition, TCO and non-depreciable working capital investments have a present value that equals initial investments, notionally:

$$\sum_{t=1}^{7} \frac{3,233}{(1+0.08)^t} + \frac{2,000}{(1+0.08)^7} = 18,000.$$

EVA as measure of created shareholder wealth equals 253 k USD, calculated as:
$$EVA_t = 3,486 - 3,233 = 253.$$
On the contrary, ROCE increases from 9.4% to 13.2% over the project life, according to the following formula:
$$ROCE_t = \frac{3,486 - ED_t^{EVA}}{Capital_{t-1}}.$$

Increasing profitability is simultaneously driven by decreases in economic depreciation and net capital. However, the increases in ROCE over the asset life are considerably lower than if EVA were not adjusted for straight-line book depreciation. Notably, returns increase over time since the cost of capital is considerably lower than the IRR; with the cost of capital being higher than the IRR, returns decrease over time.

To summarize, revised EVA implies constant levels of EVA, but built-in profitability improvements over the asset life, see Table 11.

Year	0	1	2	3	4	5	6	7
Capital, EOY	18,000	16,207	14,270	12,179	9,920	7,480	4,846	2,000
S/L depreciation		3,233	3,233	3,233	3,233	3,233	3,233	3,233
TCO		1,440	1,297	1,142	974	794	598	388
- Cost of capital		1,793	1,937	2,092	2,259	2,440	2,635	2,846
- Econ. Depreciation		3,486	3,486	3,486	3,486	3,486	3,486	3,486
"Revised" EVA		253	253	253	253	253	253	253
"Revised" ROCE		9.4%	9.6%	9.8%	10.1%	10.6%	11.4%	13.2%

Table 11   Summary of 'Revised' EVA and ROCE Calculation

Summary

To conclude, Figure 18 summarizes the above results graphically: Evidently, performance derived from depreciation-adjusted measures is less biased than from unadjusted EVA. However, depreciation-adjusted metrics differ considerably with respect to how they measure and integrate economic depreciation.

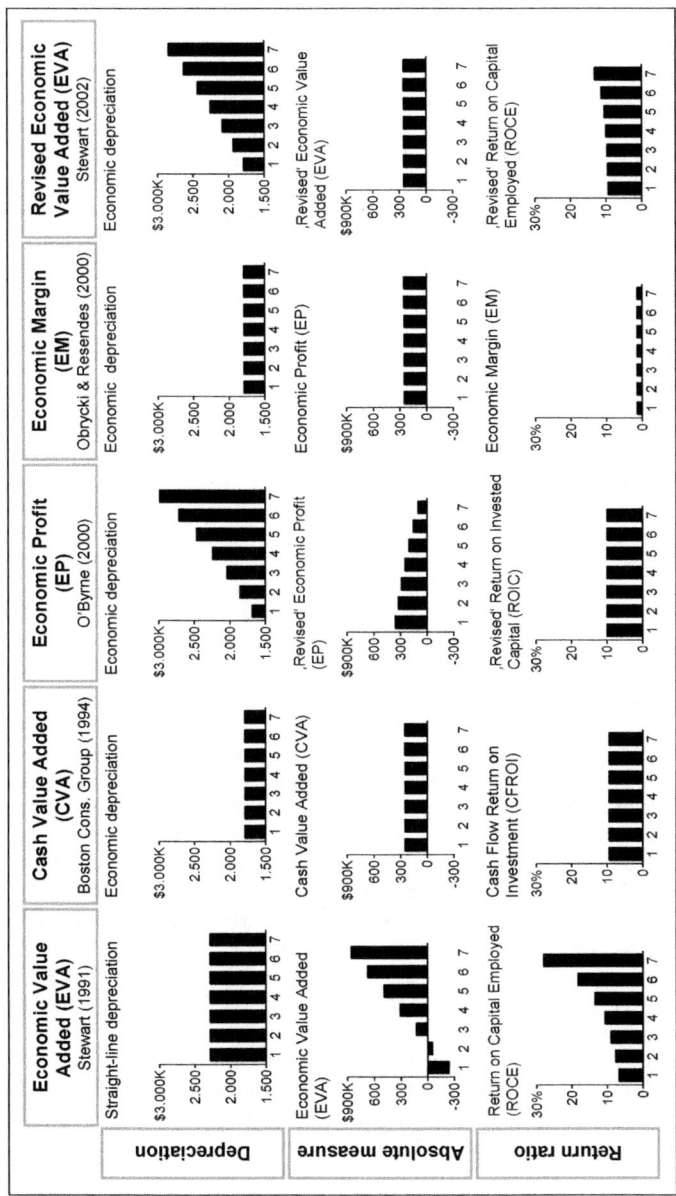

Figure 18   Outline of (Depreciation-Adjusted) VBM Metrics

In the following, the case is slightly modified to imply unequal rather than equal cash flows, to test the robustness of the measurement concepts. Provided that the net present value of the project still equals 1,319 k USD, operating cash flows before depreciation are now assumed to be 1,942 k USD for years 1 to 3 and 5,000 USD for years 4 to 7. Consequently, the IRR decreases slightly from 10.0% to 9.7%. Nevertheless, a cost of capital of 8.0% lower than the IRR still implies positive additions to shareholder value.

Figure 19 summarizes the impact of unequal cash flows on economic depreciation and absolute and relative performance. Under unequal cash flows, performance improvements of unadjusted EVA measure and the corresponding ROCE measure increase. Again, adjusted metrics differ considerably in respect of economic depreciation and performance estimates:

Following the CVA, EM or revised EVA concept, the timing of cash flows over the project time does not affect annual economic depreciation: Adjusted depreciation is either defined via a level lease payment to recover the gross depreciables or net depreciable assets :

$$ED^{CVA} = ED^{EM} = GDA \cdot \frac{r}{(1+r)^{T} - 1}$$

$$ED_{t}^{EVA} = \left(Capital_{t-1} - NDA_{t-1}\right) \cdot \frac{WACC}{(1+WACC)^{T^{*}} - 1}.$$

On the contrary, economic depreciation according to the revised EP concept is derived from changes in future discounted cash flows and, therefore, adapts to varying cash flow levels. Consequently, economic depreciation is comparatively low in initial years with low payoffs and comparatively high in final years with high payoffs.

With respect to measured value creation, CVA, EM, and revised EVA are only constant within the first or second payout period and, therefore, largely affected by changes in cash flows. Further, these measures are negative in the initial three years, reflecting insufficient cash flows in such years for generating value. On the contrary, the EP metric loses such information and does not account such difference along the project life, but avoids large changes between payout periods and negative values. However, positive EP decreases over the project life, just as under equal cash flows.

Regarding measured profitability, the revised ROIC metric succeeds in measuring constant profitability over the asset life that equals the project IRR, thus, excluding any built-in improvements from accounting depreciation and influences from changes in cash flows. CFROI and EM are affected by changes in cash flows. On the one side, CFROI is positive throughout the useful life, but it has to be compared to a minimum required return to allow statements on value addition. On the other side, EM yet allows statements on additions to shareholder wealth, showing negative values in the first periods that indicate reductions of shareholder value. The ROCE return increases under constant as well as non-constant cash flows.

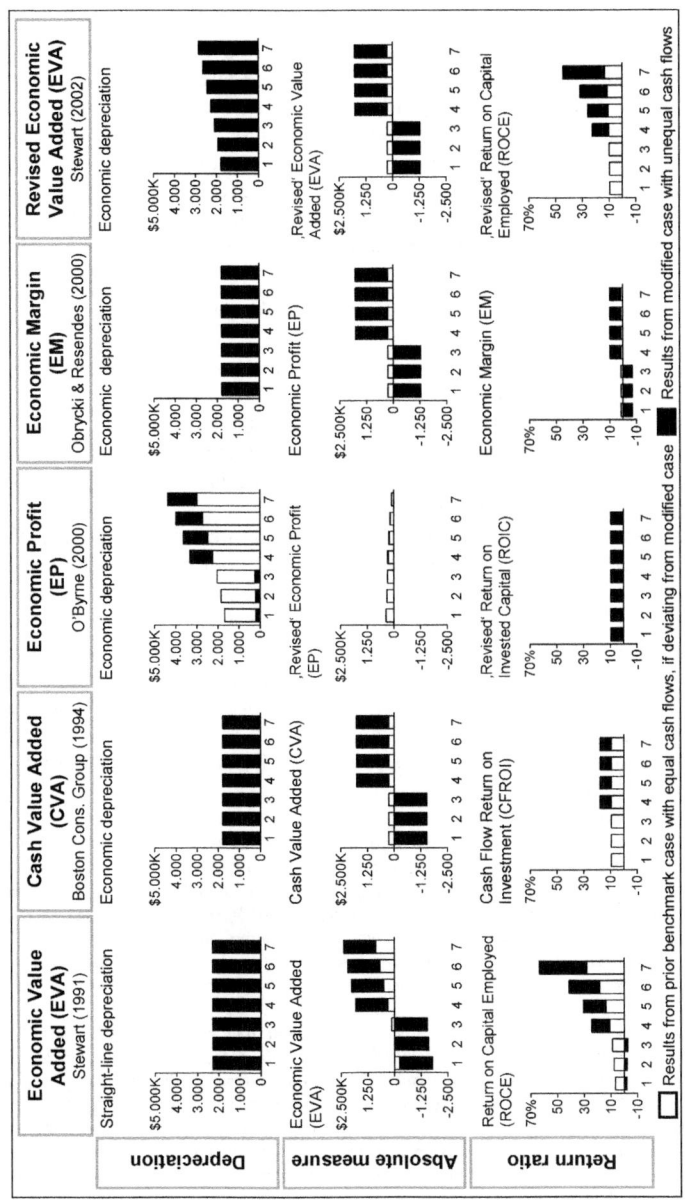

Figure 19   Robustness Test of (Depreciation-Adjusted) VBM Metrics

To conclude, all depreciation-adjusted concepts that reduce age bias from accounting depreciation offer some enhancement with respect to the initial EVA concept. The CVA and EM metric properly adjust for age bias under constant cash flows and capture the differences in the yearly value creation reflecting actual results yearly accounted by the firm. The EP concept implies less accurate indications on value addition under non-constant cash flows as it do not account negative years and, above all, actual results are biased by expectations. The revised EVA metric seems weakest, since ROCE still implies considerable depreciation-based performance improvements regardless of constant or non-constant cash flows.

3.2.2 Insufficient Advantages of Depreciation-Adjusted Metrics

Although depreciation-adjusted VBM metrics that include economic depreciation are less biased, it seems that they are not commonly used.

E.g. the annual 'Stern Stewart Performance 1000' league table ranks firms according to EVA data that includes few adjustments. EVA accounting adjustments exclude adjustments for straight-line or accelerated accounting depreciation, claiming that the depreciation adjustment is not significant.[288]

Furthermore, VBM adopters avoid the depreciation adjustment, arguing that the complexity of the depreciation adjustment and the benefit from built-in performance improvements on old investments do not support introduction of an unbiased performance measure that avoids penalizing new investments.[289] This implies that managers are reluctant to forgo relative increases of EVA from old assets, preferring to have some bias against new investment opportunities, probably also because it is supposed to be recognized (although improperly) as a kind of investment 'discipline'.

Particularly for asset-intense firms, the complexity from introducing depreciation-adjusted metrics cannot justify evading the depreciation adjustment.[290] Instead, seldom

---

[288] See Stewart (1994, p. 81).

[289] *"While the conflict between economic profit and shareholder value created by straight line depreciation can be overcome by sinking fund or positive economic depreciation, only one EVA company (that I am aware of) has ever used sinking fund depreciation for an asset that is not leased! [...] Many EVA companies have considered sinking fund depreciation, but rejected it as too complicated to justify the benefit. Their rejection of sinking fund depreciation may be influenced by the fact that the built-in EVA improvement arising from straight line depreciation on the existing asset base more than offsets the benefit of sinking fund depreciation on new assets."* (O'Byrne, 2000).

[290] *"For most companies, the straight-line depreciation of plant and equipment used in GAAP accounting works acceptably well. While straight-line depreciation doesn't attempt to match the actual economic depreciation of physical assets, the deviations from reality ordinarily are so inconsequential that they do not distort decisions. That's not true, however, for companies with significant amounts of long-lived equipment. In those cases, using straight-line depreciation in calculating EVA can create a powerful bias against investments in new equipment. That's because the EVA capital charge declines in step with the depreciated carrying value of the asset, so that old assets look much cheaper than new ones. This can make managers reluctant to replace 'cheap' old equipment with 'expensive' new gears."* (Ehrbar, 1998).

usage of depreciation-adjusted metrics may be referred to the quality of prevailing measures. In this respect, four qualitative criteria serve to evaluate each approach to reduce bias from accounting depreciation, namely: explicit charge for economic depreciation charge, unbiased measure of value addition, unbiased measure of profitability, and objective and 'easy' calculation, as follows.

First, subtraction of adjusted depreciation from earnings as an explicit depreciation charge ensures that the metric is easily understood by operating managers and investors. On the contrary, a combined charge for the cost of capital and depreciation adds complexity and imposes communication hurdles. Second, the absolute measure of value added should be distorted by age bias and by other discretional factors as less as possible. Third, the corresponding relative profitability measure should similarly not be influenced by the age of the asset and by other bias. Fourth, all measures, except revised EVA, imply some complexity and subjectivity in their calculation methodology.

The qualitative characteristics of the VBM metrics can be assessed as follows (see Figure 20):

- Explicit charge for economic depreciation charge: The CVA and revised EP metrics, explicitly deducting economic depreciation, dominate the EM and revised EVA metrics, which include charges for economic depreciation in the combined 'annuity level payment'.[291].

- Unbiased measure of value addition: The CVA, EM and revised EVA concept involve unbiased measures, regardless of constant or unequal cash flows. EP relatively decreases over the asset life, although value added is stable over time. .

- Unbiased measure of profitability: The return measure ROCE exhibits a relative increase over time, although no additional value adding activities have been taken place. The CFROI and EM measure are not distorted by age bias under constant cash flows and they somehow reflect changes in value creation associated to changes in cash flows. ROIC represents the more stable profitability measure with respect to unequal cash flows but this has to be considered a negative aspect as it imply a distortion in respect of the actual result accounted by the company and it is, in fact, heavily biased by expectations on future cash flows and IRR.

- Objective and 'easy' calculation: EP is heavily biased by expectations (IRR and cash flows). Therefore, EP does not properly account the actual results of the firm in terms of value creation actually generated (rather than expected). The CVA method is commonly known to be too complex, primarily due to inflation adjustments and the IRR computation of CFROI, as mentioned before.[292] Equally, the EM concept, merging the CVA and EVA metric, cannot be considered simple and objective.

---

[291] Combined charges of EM and EVA are called 'capital charge' and 'total cost of ownership', respectively.

[292] See sections 1.1.1 and 2.2.2.

	Cash Value Added (CVA)	'Revised' Economic Profit (EP)	Economic Margin (EM)	'Revised' Economic Value Added (EVA)
Explicit depreciation charge	+	+	-	-
Unbiased measure of value addition	+	-	+	+
Unbiased measure of profitability	+	+	+	-
Objective and 'easy' calculation	-	-	-	+

Figure 20   Assessment of Depreciation-Adjusted VBM Metrics

Consequently, the CVA concept stands out, fulfilling three out of four imposed qualitative criteria. At the same time, this analysis shows the need for new depreciation-adjusted VBM metrics, as introduced in the following section.

3.3   Introduction of New Value-Based Metrics Using Economic Depreciation

3.3.1   Definition of CRI and CEVA as New Depreciation-Adjusted Metrics

It is important to avoid age bias, especially in asset-intense industries that invest heavily in long-lived physical assets, such as real estate, media, telecom, hotel chains, and transportation. Without adjusting for accounting depreciation, management may abstain from essential investments in new assets to sustain value creation and investors may misunderstand value creation of single firms. On the other side, it has to be noted that in 'mature' companies (with a relatively low investment rate) EVA and RI metrics generally provide higher profitability ratios and annual value creations with respect to the relevant values accounted offsetting the age bias.

Diverse rationales may explain why common depreciation-adjusted VBM metrics are rarely used: e.g., adopters may prefer avoiding inflation adjustments (and therefore CVA), avoiding bias derived from expectations (and therefore revised EP or ROIC), avoiding complex and arbitrary cost of capital computations (and therefore CFROI), and avoiding opaque combined charges for capital and depreciation (and therefore EM and revised EVA).[293]

---

[293] E.g., Madden (1999) adjusts CFROI (and thus CVA) for inflation and applies a market-derived cost of capital based on forecasts for a global portfolio of firms and average fade rates for mini-aggregates of similar firms.

Therefore, are here newly introduced two metrics, 'Cash Residual Income' (CRI) and 'Cash Economic Value Added' (CEVA), that avoid bias of the revised EP and EVA concepts, inflation adjustments of the CVA concept, and implicit depreciation charges as included in the EM concept. Both measures deduct economic depreciation and refer to gross assets, avoiding relative performance increases and underestimation of value creation in the early years of significant investments. CRI and CEVA are derived from two prevalent comprehensive VBM measures, namely RI and EVA, which are converted onto a gross earnings and asset basis (before non-cash depreciation and amortization), as follows.

Cash Residual Income (CRI)

'Cash Residual Income' (CRI) aims to overcome age bias inherent in basic RI. RI represents the most basic VBM metric and is defined as $RI_t = NE_t^{BI} - WACC_t \cdot NA_{t-1}$, where NE^{BI} = net earnings before interest expenses, $WACC$ = weighted average cost of capital, and NA = book value of net assets.[294] Instead, CRI is defined as cash earnings, CE^{BI}, less economic depreciation, ED, and a capital charge based on gross assets, GA, and the weighted average cost of capital, $WACC$, notationally:

$$CRI_t = CE_t^{BI} - ED_t - WACC_t \cdot GA_{t-1}.$$

Importantly, CE^{BI} shall exclude charges for interest expense and depreciation and amortization. ED is defined as annuity that amounts to the acquisition cost of depreciable assets at the end of the useful life.[295] Finally, GA shall include accumulated depreciation and amortization.

CRI also can be written as spread on gross assets, based on the corresponding return measure 'Return on Gross Assets', $ROGA$, as follows:

$$CRI_t = \left(\frac{CE_t^{BI} - ED_t}{GA_{t-1}} - WACC_t \right) \cdot GA_{t-1} = (ROGA_t - WACC_t) \cdot GA_{t-1}.$$

---

[294] For a definition of RI, see section 2.2.2.

[295] Economic depreciation is calculated based on the annuity formula: $ED_t = GDA_{t-1} \cdot \dfrac{WACC}{(1+WACC)^T - 1}$, where GDA = gross depreciating assets and T = average useful life of assets. If the average asset life is unknown, it may be approximated as $T = \dfrac{GDA_{t-1}}{Dep_t}$, where Dep = book depreciation and amortization.

Cash Economic Value Added (CEVA)

'Cash Economic Value Added' (CEVA) aims to overcome age bias included in EVA, while continuing to adjust for other distortions yet accounted for under EVA. Contrary to RI, EVA includes several accounting adjustments to net earnings and net capital and is defined as $EVA_t = NOPAT_t - WACC_t \cdot Capital_{t-1}$.[296] As a depreciation-adjusted substitute, CEVA represents adjusted 'cash' earnings (cash operating profits after taxes), $COPAT$, less economic depreciation, ED,[297] and a capital charge based on adjusted gross assets, $Gross\ Capital$, and the weighted average cost of capital, $WACC$, notationally:

$$CEVA_t = COPAT_t - ED_t - WACC_t \cdot Gross\ Capital_{t-1}$$

Importantly, COPAT shall exclude charges for interest expense and depreciation and amortization and additionally but include any EVA accounting adjustments to earnings. Consistently, Gross Capital shall include accumulated depreciation and amortization and any EVA accounting adjustments to capital.

CEVA can be rewritten as spread on adjusted gross capital, defining the implied return measure 'Cash Return on Capital Employed', $CROCE$, as COPAT after economic depreciation divided by beginning gross capital, as follows:

$$CEVA_t = \left(\frac{COPAT_t - ED_t}{Gross\ Capital_{t-1}} - WACC_t \right) \cdot GrossCapital_{t-1} =$$

$$= (CROCE_t - WACC_t) \cdot Gross\ Capital_{t-1}$$

To avoid double counts, COPAT and Gross Capital calculation requires attention with respect to goodwill amortization and costs for debt financing. Both expenses are yet considered within EVA accounting adjustments, i.e., after-tax interest expenses and goodwill amortization and impairment charges are added back to earnings and accumulated goodwill amortization and unrecorded goodwill are added back to net capital.[298]

---

[296] For a definition of EVA, see section 2.2.2 and 2.3.1. Stewart just refers to adjusted net capital as 'capital'. While not specifically termed net capital, 'capital' excludes cumulative depreciation and amortization.

[297] As for CRI, economic depreciation is calculated as annuity: $ED_t = GDA_{t-1} \cdot \dfrac{WACC}{(1+WACC)^T - 1}$, where GDA = gross depreciating assets and T = average useful life of assets (which can be approximated as $T = \dfrac{GDA_{t-1}}{Dep_t}$, where Dep = book depreciation and amortization).

[298] See Stewart (1991, pp. 78, 114-115, 177).

Reconciliation of CRI and CEVA with Corporate Value

Most important, CRI theoretically reconciles with the value of the firm (and likewise CEVA). Namely, corporate value equals to the sum of gross assets and discounted future CRIs, as shown by the following mathematical proof (assuming that WACC is constant).

Shown in the first line, the discounted cash flow model is equivalent to the residual income valuation relationship.[299] Notably, the relationship is still valid applying any change to the depreciation method, as long as the capital is consistently accounted.[300] On this basis, in the second line, book depreciation is substituted with economic depreciation starting at time t, which requires holding net capital at time t constant (being the economic depreciation equivalent to the annuity matching the present value of the complete depreciation of assets at the end of asset life;[301] additionally, accumulated book depreciation and amortization, ADA, at time t is added to and subtracted from net capital. Finally, the equation is rewritten to express corporate value via gross assets and future CRIs.[302]

$$CV_t = \sum_{j=1}^{\infty} \frac{FCF_{t+j}}{(1+WACC)^j} = NA_t + \sum_{j=1}^{\infty} \frac{RI_{t+j}}{(1+WACC)^j} = NA_t + \sum_{j=1}^{\infty} \frac{NE_{t+j}^{BI} - WACC \cdot NA_{t+j-1}}{(1+WACC)^j} =$$

$$= NA_t + \sum_{j=1}^{\infty} \frac{\left(NE_{t+j}^{BI} + Dep_{t+j}\right) - ED_{t+j} - WACC \cdot \left(NA_t + ADA_t - ADA_t\right)}{(1+WACC)^j} =$$

$$= NA_t + \sum_{j=1}^{\infty} \frac{CE_{t+j}^{BI} - ED_{t+j} - WACC \cdot GA_t}{(1+WACC)^j} + \sum_{j=1}^{\infty} \frac{WACC \cdot ADA_t}{(1+WACC)^j} =$$

$$= NA_t + \sum_{j=1}^{\infty} \frac{CRI_{t+j}}{(1+WACC)^j} + ADA_t =$$

$$= GA_t + \sum_{j=1}^{\infty} \frac{CRI_{t+j}}{(1+WACC)^j}$$

q.e.d.

---

[299] See section 2.2.2 and 4.2.2.

[300] This independence is proven in the following subsection.

[301] The relation implies a depreciation scheme equal to zero from the time $t+1$ until the end of each asset life, T, and a complete depreciation of the whole asset value (write-off) at T; the relevant capital charges of CRI and CEVA have to be therefore referred to constant net assets, equal to NA_t; on the other side, the present value of the complete depreciation of the assets NA_t at the end of the period T is equal to the present value of the sum of EDs, being ED calculated as the annuity of the amount NA_t at the time T.

[302] The reformulation uses the following equivalences: $\sum_{j=1}^{\infty} \frac{WACC \cdot a}{(1+WACC)^j} = a$ and $NA_t + ADA_t = GA_t$.

The proof above is assumed to be equally valid for CEVA (that represents a residual income metric), i.e. corporate value shall equal the sum of gross assets and the present value of future CEVAs:

$$CV_t = NA_t + \sum_{j=1}^{\infty} \frac{EVA_{t+j}}{(1+WACC)^j}\bigg|_t = GA_t + \sum_{j=1}^{\infty} \frac{CEVA_{t+j}}{(1+WACC)^j}.$$

The reconciliation shows that RI (likewise EVA) and CRI (likewise CEVA) measures refer to different benchmarks: While RI (and EVA) describes value added to net capital, CRI (and CEVA) quantifies value added to gross capital. Since both measures reconcile with corporate value, (cumulative) CRI or CEVA is lower than (cumulative) RI or EVA, respectively.

Even if RONA, ROCE, ROGA and CROCE are all immediately comparable with WACC, it may be wrongly assumed that they are also comparable among each other. All such profitability ratios provide an effective percentage-based indicator of value performance. In particular, it is for each of them the spread versus WACC that is the relative measure of value creation (or destruction). But again, it is important to have in mind that for ROGA and CROCE such value creation is referred to the value of gross assets, whilst the traditionally used RONA and ROCE refer to the value of net assets.

The Underlying Concept of the New Metrics

The proof above assumed that a change to the depreciation method, if consistently accounted, does not affect corporate value. This independence can be demonstrated in two ways.

First, a logical demonstration combines the following: being corporate value at time t equal to the present value of expected free cash flows and, therefore, not depending on the future depreciation method applied, and being also corporate value at time t equal to net assets and the present value of expected residual incomes, being NA_{t-1} clearly independent from future depreciation methods, also the sum of future discounted residual incomes must be independent from the future depreciation method applied.

Second, a more analytical demonstration, from which can be also derived as a particular case the equivalence of corporate value with future discounted free cash flows, is the following: To demonstrate the independence of the RI valuation relationship it has to be demonstrated the equivalence of the present value of two RI series where different depreciation methods are applied; in this respect, it is sufficient to demonstrate the above when postponing an indefinite depreciation charge, Dep, for a certain number of years, as any accounting depreciation method has to provide the same accumulated depreciation, equivalent to the asset value. To begin with, the demonstration shows that postponing from the year K to the following year $K+1$ the account of any depreciation charge, Dep, the present value of expected RI do not change.

At this scope, being RI_K the series of the RIs before the postponement of the depreciation (and amortization) charge, Dep, and RI_K^* the series of the RIs after the postponement, it is

$NE_K^* = NE_K + Dep$, $NE_{K+1}^* = NE_{K+1} - Dep$, and $NA_{K-1}^* = NA_{K-1}$, $NA_K^* = NA_K + Dep$, $RI_i^* = RI_i$ for every i different from K and $K+1$, from which:

$$CV_t = \sum_{j=1}^{\infty} \frac{RI_{t+j}}{(1+WACC)^j} =$$

$$= \sum_{j=1}^{K-1} \frac{NE_{t+j} - WACC \cdot NA_{t+j-1}}{(1+WACC)^j} + \frac{NE_K - WACC \cdot NA_{K-1}}{(1+WACC)^K} + \frac{NE_{K+1} - WACC \cdot NA_K}{(1+WACC)^{K+1}} +$$

$$+ \sum_{j=K+2}^{\infty} \frac{NE_{t+j} - WACC \cdot NA_{t+j-1}}{(1+WACC)^j} =$$

$$= \sum_{j=1}^{K-1} \frac{RI_{t+j}}{(1+WACC)^j} + \frac{NE_K^* - Dep - WACC \cdot NA_{K-1}}{(1+WACC)^K} +$$

$$+ \frac{NE_{K+1}^* + Dep - WACC \cdot \left(NA_K^* - Dep\right)}{(1+WACC)^{K+1}} + \sum_{j=K+2}^{\infty} \frac{RI_{t+j}}{(1+WACC)^j} =$$

$$= \sum_{j=1}^{K-1} \frac{RI_{t+j}}{(1+WACC)^j} + \frac{RI_K^*}{(1+WACC)^K} - \frac{Dep}{(1+WACC)^K} + \frac{Dep \cdot (1+WACC)}{(1+WACC)^{K+1}} +$$

$$+ \frac{RI_{K+1}^*}{(1+WACC)^{K+1}} + \sum_{j=K+2}^{\infty} \frac{RI_{t+j}}{(1+WACC)^j} =$$

$$= \sum_{j=1}^{K-1} \frac{RI_{t+j}^*}{(1+WACC)^j} + \frac{RI_K^*}{(1+WACC)^K} + \frac{RI_{K+1}^*}{(1+WACC)^{K+1}} + \sum_{j=K+2}^{\infty} \frac{RI_{t+j}^*}{(1+WACC)^j} =$$

$$= \sum_{j=1}^{\infty} \frac{RI_{t+j}^*}{(1+WACC)^j}$$

q.e.d.

Repeating the above for n times (and also for negative values of Dep), it can be derived that the equivalence $RI_i^* = RI_i$ still remains valid postponing or anticipating any depreciation charge, Dep, for whatsoever number of years n, from which is demonstrated the independence of the result of the present value of the RI series from the applied depreciation method.

3.3.2   Advantages of CRI and CEVA versus Common VBM Metrics

In a nutshell CRI and CEVA intend to capture the key features as well as the methdological advantages of the RI and EVA metrics, with the better performance pattern of the CVA metric overcoming the problem of the age bias.

In particular, similarly to RI and EVA, the CRI and CEVA provide various advantages versus other metrics:

- Solid definition, since theoretically reconciling with corporate value

- Transparent and easily understandable formulation that is directly linked to accounting results

- Value creation clearly referred to each single accounted period on the basis of yearly accounting results

- No additional arbitrary assumptions, parameter or expectation bias, apart from the discretion (if any) applicable within the applied accounting principles, affecting the accounted results

- Easy and consistent deployment within the overall organization based on the additivity of single business unit results and a corresponding value driver tree (value drivers derived from components that reach across organizational levels)

- Profitability ratios (ROGA and CROCE) comparable with a clear, well known and already broadly applied profitability value threshold, namely the weighted average cost of capital (WACC)

Additionally, CRI and CEVA provide the same results as CVA in terms of offsetting age bias, providing the following benefits:

- Stable yearly value creation undistorted by built-in performance improvements due to book depreciation

- Stable profitability ratios (ROGA and CROCE) that are not affected by understatements in early years of the asset life and overstatements in later years of the asset life

- Transparent and easily understandable revised (economic) depreciation charge

- No internal rate of return or free cash flow expectations to estimate economic depreciation

3.4 Summary

Net residual income measures that are defined after accounting depreciation, such as RI and EVA, imply built-in annual performance improvements over the asset life, referred to as age bias. Such bias emerges under various US-GAAP depreciation methods, and implicates a biased investment policy and biased portfolio selection.

To overcome age bias, depreciation-adjusted metrics, such as CVA, EM, revised EP, and revised EVA, replace book depreciation with economic depreciation. Methodologically, the CVA metric stands out: Both the CVA measure and the corresponding return measure CFROI remove age bias and economic depreciation is deducted via an explicit depreciation charge. However, the CVA implies complex and subjective adjustments.

Overall, such depreciation-adjusted methods are not commonly used, presumably due to their complex and not intuitive calculation that in some cases is also affected by arbitrary components. For these reasons, this study develops and describes two new depreciation-adjusted metrics: CRI and CEVA. The first is a basic depreciation-adjusted VBM metric that avoids any complexity, whilst CEVA, relating to the prominent EVA measure, is also adjusted for several other accounting distortions.

It is commonly known that RI (and EVA) is congruent with the discounted cash flow valuation theory. This means that corporate value, assumed equal to the present value of expected future free cash flows, can be also obtained by summing up net assets and the present value of expected future RIs (or net capital and the present value of expected EVAs). As this principle constitutes a fundamental property for both RI and EVA metrics, it is also important that it also can be applied to the new CRI and CEVA metrics. This study provides a demonstration of such value equivalence.

Another important property of the CRI and CEVA metrics (likewise for RI and EVA) is that the correspondent new profitability indexes, 'Return On Gross Assets' (ROGA) and 'Cash Return on Capital Employed' (CROCE), when compared with the cost of capital (i.e. the traditional WACC) still indicate whether shareholder value is added (positive spread) or destroyed (negative spread).

At this point, one more issue has to be considered for completing the interpretation framework for these value metrics: Under VBM metrics, quantified value creation always refers to a certain value level that depends on the specific metric. In this respect, whilst RI and EVA provide an indication of value creation with respect to the book value of net assets, CRI and CEVA provide a measure of value creation with respect to the book value of gross assets.

4 Applied Methods for the Empirical Research

This chapter defines the study design that provides the methodological basis of the subsequent study.

First, a review of value-relevance studies on value-based metrics demonstrates the underlying research gap. Accordingly, four research hypotheses incorporate assumptions on ordinary traditional and value-based performance metrics as well as suppositions on newly defined metrics. The overall study design integrates these hypotheses, based on regression models that enhance and conflate prevailing approaches and independently defined variables that ensure replicability of the study.

4.1 Review of Prior Value-Relevance Research

VBM requires that the performance metric not only reflects creation of shareholder wealth theoretically but also explains shareholder returns in practice.[303] Respective empirical studies are part of value-relevance research that aims to determine whether an accounting number is useful for valuing the firm by testing the association of the accounting value measures with stock price developments. Four different basic types of analysis are commonly used to examine value-relevance of a performance metric: measurement analysis and analysis of relative, incremental, and marginal information content.[304]

This study examines the relative and incremental information content of VBM metrics:[305] First, relative information content studies compare the association of bottom-line measures with levels or changes in market prices. The higher the level of R^2, the more value-relevant a metric is. Second, incremental association studies assess

---

[303] *"The interest of shareholders must be the focus of any performance-measurement system. The success of a performance-measurement system at tracking and consistently reflecting shareholders' interests demonstrates how well it achieves this objective."* (Knight, 1998, p. 203).

[304] See Holthausen and Watts (2001).

[305] The research objective of this study is neither to find a suitable measure for stock valuation (as the objective of measurement studies) nor to explain short-term changes in stock prices after the release of value-based information (as the motivation of marginal information content studies). Measurement studies incrementally test the measurement error, with which an accounting number estimates an input variable of an equity valuation model. Marginal information content studies examine whether the release of specific accounting numbers is associated with changes in stock value, typically using an event study approach.

whether a unique additional information component aids in explaining firm value or stock returns. A metric is value-relevant if the estimated regression coefficient of its incremental information component is significantly different from zero.

The following review discusses value-relevance studies that include at least one value-based accounting measure, distinguishing between studies examining the relationship between firm performance with either market value or stock returns.

4.1.1 Empirical Studies Relating Accounting Performances to Market Value

Commonly, empirical studies relate corporate performance measures to the absolute market value of the stock, since financial theory supports associations of cash flows and residual income with stock prices.[306] Usually, these studies examine relative information content of competing performance measures based on a consistent linear regression model.[307]

To emphasize the development over time, Table 12 provides a chronological overview on studies examining associations with market value. Next to the results, the table provides information on five qualitative criteria, defined as follows:

- Sample size: The panel data shall include firm-year observations from a sufficient number of firms and years.

- Sector (country): Firm-year observations shall originate from stable, mature markets, not primarily driven by external factors or long-term expectations.

- Scaling: Dependent and independent variables shall be consistently scaled by book value to avoid that association is primarily driven by firm size.

- Regression on single firm-years: Association analysis shall be based on regression rather than correlation statistics. Furthermore, performance from single firms and years rather than long-term performance from equity portfolios shall be the unit of analysis and one regression model shall be consistently applied for any independent variable.

- Data source: Independently calculated data from public databases shall be preferred to data provided by management consultancies, such as Stern Stewart or Boston Consulting Group.

---

[306] See sections 2.2.2 and 4.2.2.
[307] Except O'Byrne (1996) and Tsuji (2006) that relate perpetual performance level rather than current performance level to market value.

Author (Year)	Sample size	Sector (Country)	Scaling *	Model **	Data source	Outperforming measure
Finegan (1991)	467 firms ('84-'85 / '87-'88)	Industrials (U.S.)			Stern Stewart	EVA (r_S= 60%)
Stewart (1991)	613 firms ('84/'85,'87/'88)	Industrials (U.S.)			Stern Stewart	N/A
O'Byrne (1996)	870 firms (1985-1993)	Industrials (U.S.)	Yes		Stern Stewart	EVA (R^2 = 56%)
Biddle et al. (1997)	773 firms (1983-1994)	Industrials (U.S.)	Yes	Yes	Stern Stewart	Earnings (R^2 = 53%)
Kramer & Pushner (1997)	1,000 firms (1982-1992)	Industrials (U.S.)	Yes	Yes	Stern Stewart	Earnings (R^2 = 6%)
Clinton & Chen (1998)	325 firms (1991-1995)	Industrials (U.S.)			Stern Stewart	Earnings (r_S = 56%)
Heidorn et al. (2001)	91 firms (1992-1999)	7 sectors (Europe)		Yes	Annual reports, Deutsche Bank	Earnings (R^2 = 56%)
Fernández (2001)	582 firms (1988-1997)	Industrials (U.S.)			Stern Stewart	Δ Earnings (r_S = 23%)
Fernández (2002)	28 firms (1992-1998)	Industrials (Spain)			Stern Stewart	EVA (r_S = 56%)
Garg & Singh (2004)	50 firms (1998-2002)	Industrials (India)			Annual reports, BSE directory, CIME, Hisar	Earnings (R^2 = 15%)
Tsuji (2006)	561 firms (1982-2002)	All sectors (Japan)	Yes	Yes	AMSUS, Japan Securitites Research Institute	Cash flow (R^2 = 77%)
Ramana (2007)	50 firms (1995-2006)	All sectors (India)	Yes	Yes	CMIE Prowess database	Earnings (R^2 = 80%)

Note: Considerably differing levels of R^2 result from the individual model specifications.
* Firm performance and market value consistently scaled by book value.
** Regression model based on single firm-years and consistently applied.

Table 12   Studies Relating Performance Measures to Firm Value

Analysis on EVA has been initiated by Stern Stewart in-house studies, which under-lined the superiority of EVA in explaining firm value.

In 1991, Stewart highlighted the validity of EVA by stating that the 1984/1985 to 1987/1988 change in EVA scaled by initial capital explains 97% of the respective change in MVA scaled by initial capital, significant at the 1% level. Equally, 1987/1988 levels of EVA explain 96% of 1988 level of MVA.[308] While demonstrating that EVA is highly correlated with MVA, Stewart omits to test competing metrics.

Complementary, Finegan compares EVA with alternative performance measures, finding that EVA is predominant to traditional performance measures: The 1987/1988 level of EVA explains 60% of the 1988 level of MVA, whereas the 1988 return on capital, the 1984-1988 growth in cash flow, and earnings per share reflect only 47%, 20%, and 10% of the 1988 level of MVA, respectively.[309]

However, the study designs of these two studies involve several issues: First, Stewart excluded poorly performing firms, assuming that investors expect reverse positive future earnings of recent investments, not captured by EVA.[310] Finegan excluded the top and bottom quartiles of the initial 900 largest industrial U.S. firms based on 1988 MVA, to focus on firms with moderate market valuation, and used differing time intervals across measures. Moreover, changes based on average data of two consecutive fiscal years avoid short-term distortions in earnings and market valuation and prevent continuous time-series analysis. Additionally, using groups of 25 firms assembled by average EVA instead of single firm observations reduces valuation differences across individual firms. Finally, both studies are correlation rather than regression analyses, therefore, preventing any conclusions on causality.

In 1996, another in-house study by O'Byrne provided further support for the superiority of EVA based on a regression analysis:[311] Accordingly, EVA explains 56% of MVA, whereas NOPAT and free cash flows (FCF) explain 17% and 0% of MVA, respectively.

Although all variables are consistently deflated by beginning capital, the overall study design is biased: Contrary to the earnings model, the EVA model does not requires a zero intercept, capitalizes EVA by the cost of capital, allows the coefficients for positive and negative values of EVA to differ, and adds industry coefficients and the natural logarithm of the book value as an additional size variable.[312]

To summarize, Stern Stewart research consistently provided evidence on the outperformance of EVA, however, lacking a profound study design.

---

[308] See Stewart (1991, pp. 215-218).
[309] See Finegan (1991).
[310] Firms with that earn at least 2.5% less than their cost of capital and minimum capital growth of 25% p.a. are identified as poorly performing.
[311] See O'Byrne (1996).
[312] O'Byrne assumes that a corporate size variable reflects the fact that productivity gains of larger corporations are less likely due to increasing bureaucratic costs and decreasing marginal scale effects and experience. Industry coefficients reflect that investors' evaluate stocks based on the respective industry of the firm.

Given the earnings myopia of financial markets, academicians reexamined non-intuitive dominance of EVA over earnings.

In 1997, Biddle et al. (1997), applying a consistent regression model across variables, relativize O'Byrne's claims on EVA:[313] Earnings are significantly higher associated with firm value than EVA and NOPAT, explaining 53%, 50% and 49% of variations in market value, respectively. Furthermore, EVA does not significantly outperform NOPAT. The underlying regression model accounts for industries, sign of performance, and firm size and deflates firm performance by beginning capital.

In 1997, Kramer and Pushner reconfirm the superiority of earnings:[314] NOPAT is predominant to EVA, explaining 18% and 10% of MVA, respectively. After elimination of firm-years with standard errors above or below two standard deviations from the mean, the association of current-year EVA with MVA increases from 10% to 30%.[315] Discrepancies to the level of adjusted R^2s presented by Biddle et al. result from the omission of additional variables. However, it seems that associations are primarily driven by size: After scaling all variables by beginning capital, NOPAT and EVA explain only 6% and 5% of MVA, respectively.

Then, several studies reexamined the value-relevance of EVA, applying a rather simplistic study design. In 1998, Chen and Clinton provide a correlation analysis on accounting measures and annual closing stock prices:[316] While EVA and RI per share do not significantly correlate with equity prices, NOPAT per share (56%) and operating cash flow per share (55%) significantly correlate. In 2001, Fernández finds that changes in NOPAT (23%) are more highly correlated with changes in MVA than changes in EVA (18%).[317]

Overall, independent regression studies show that earnings dominate EVA, applying a consistent study design.

To conclude, independent regression and correlation analyses consistently find that earnings are more value-relevant than EVA, contrary to in-house findings of Stern Stewart & Co..

So far, research examined the value-relevance of EVA for U.S. firms based on readily available data on the 1,000 largest U.S. firms, published and promoted by Stern Stewart & Company. Subsequently, studies examined the information content of EVA outside the United States.

In 2001, Heidorn et al. concluded that EVA is more-value relevant than earnings for European firms:[318] Results show an association of EVA divided by book value with MVA divided by book of 56%, which increases to 61% after excluding firms with scaled EVA values of less than 2.5%. Instead, return on invested capital divided by the cost of capital explains only 48% of the variation in scaled firm value. However, it

---

[313] See Biddle et al. (1997, p. 331).
[314] See Kramer and Pushner (1997).
[315] See Kramer and Pushner (1997, p. 43).
[316] See Chen and Clinton (1998).
[317] See Fernández (2001).
[318] See Heidorn, Siebrecht, and Klein (2001).

seems questionable to compare measures with differing scaling factors. Additionally, findings relate to a relatively small number of 728 observations and EVA data include a limited number of standard accounting adjustments.

In 2002, Fernández finds for 28 Spanish firms that the correlation of EVA (56%) is slightly higher than that of EP (50%), profit after taxes (44%), and equity cash flow (39%) with shareholder value added. [319] However, the rather limited sample and insufficient disclosure of the definition of variables suggest caution when stressing these results.

In 2004, Garg and Singh examine statistical associations in India, showing that NOPAT dominates explaining 15% of MVA, while EVA, profit after taxes divided by net capital, net operating profit divided by equity capital, and earnings-per-share explain 11%, 1%, 0.1%, and -2% of MVA. [320] However, EVA equals in fact RI, since EVA calculation omits accounting adjustments.

In 2006, Tsuji, examining Japanese firms, finds that earnings and cash flows dominate EVA: [321] The association of EVA (15%) with MVA is considerably lower than the association of cash flow (77%) and profit after tax (46%). However, EVA is divided by the cost of capital, while earnings and cash flow not.

In 2007, Ramana examines performance measures for 50 Indian firms, finding that earnings outperform EVA: [322] NOPAT explains 80% of MVA scaled by capital. Still highly-significant, after-tax profit and cash profit explain about 29% of scaled MVA. EVA does not explain scaled MVA significantly ($R^2 = 0.2\%$). However, EVA is independently calculated to actually equal basic RI. Further, results are to be considered carefully, since the sample of 50 firms is quite small and includes financial institutions, which publish financial statements not comparable to non-financial firms.

To conclude, non-U.S. studies provide ambiguous results, however, including major deficiencies in the study design and data.

All in all, findings overall refute EVA's dominance, while only studies from Biddle et al. and Kramer and Pushner use a robust data basis and study design. Moreover, evidence shows that the relationship between EVA and the firm's market value depends on the stage of the business cycle, being more stable in contraction rather than expansion periods, [323] while not depending on the sectors' capital intensity. [324]

---

[319] See Fernández (2002a).

[320] See Singh and Garg (2004).

[321] See Tsuji (2006); noteworthy, negative adjusted R^2s values imply that the sum of squares of the distances between the data points and the mean of all Y values is smaller than the sum of the vertical distances of each point from the best-fit curve, i.e., a horizontal line provides better estimates than the best-fit curve.

[322] See Ramana (2007).

[323] See results from Kim, Ahn and Yun (2004, Spring) based on a sample of 1,109 firms included in the Stern Stewart ranking database for the years 1990 to 1995.

[324] See results from 972 firms referring to 53 industries, as included in the Stern Stewart ranking database, for the years 1978 to 1996 (Kramer & Peters, 2001); the authors approximated the level of capital intensity of each industry by the median fixed asset turnover ratio.

4.1.2 Empirical Studies Relating Accounting Performances to Stock Returns

Other empirical studies relate performance measures to stock returns. Table 13 presents a chronological overview that includes qualitative criteria introduced above. With respect to the regression model, return studies shall preferably apply an expectancy model, such as the level and changes or the one-lag model, to approximate unexpected firm performance and deflate by the market value of equity, to decrease heteroscedasticity.[325]

Initially, researchers conducted correlation analyses based on U.S. firms.

In 1996, Peterson and Peterson refute expectations on higher correlations of more sophisticated value-based measures than those of most naïve traditional measures: [326] Average correlation of traditional measures, such as net earnings on equity and cash flow returns on assets, is 17% and 14%, respectively, while that of ROCE and EVA is 13% and 11%. Moreover, after dividing annual samples into firms with positive and negative EVA, positive EVA firms outperform significantly only in two out of five years. While correlations with total shareholder returns and size-adjusted returns are comparable, correlations with market-adjusted returns are relatively low, on average 1%, resulting either from a wrong estimates of market expectations or from prevailing valuation mechanisms that do not recognize prior return expectations.

In 1997, the results from Chen and Dodd indicate that EVA or RI explain stock prices better: [327] Regressing groups of variables based on 10-year averages, EVA variables (41%) and RI variables (36%) explain a higher share of returns than traditional earnings variables (11%).[328] The relatively low R^2 of earnings variables may result from the earnings subgroup contain only three return variables, while the RI and EVA subgroups both consist both of four variables and include a growth and change variable.

In 1998, Chen and Clinton find that cash flows and earnings outperform value-based measures: Operating cash flow (31%) and NOPAT (27%) gain the highest positive correlation with returns, while EVA is negatively and extremely low correlated and RI is insignificantly correlated with returns.[329]

To summarize, correlation studies provide ambiguous results, but indicate that traditional measures outperform EVA.

---

[325] The level and changes and one-lag model are introduced in section 4.2.2.

[326] See Peterson and Peterson (1996); the authors estimate market returns using a market model and applying parameters from prior 60 monthly returns; size-adjusted return is the excess return with respect to the equally weighted average return of the size decile of all NYSE firms; overall, the article primarily provides a guideline how to calculate EVA by applying six standard adjustments and an illustrative case study.

[327] See Chen and Dodd (1997); RI is computed as EVA less equity equivalents, thus, common calculation principles of EVA and RI variables may explain similar association with returns.

[328] Results on individual measures are only provided for EVA variables: Stand-alone EVA is leading the EVA variables: EVA per share explains 11% of raw returns, whereas the EVA capital growth explains 8%, the change in EVA 7% and the EVA spread 5% of raw returns.

[329] See Chen and Clinton (1998).

Author (Year)	Sample size	Sector (Country)	Scaling *	Model **	Data source	Outperforming measure
Olsen (1996)	1122 firms (1994-1995)	Industrials (U.S.)			BCG, Value Line	TBR ($r_S = 40\%$)
Peterson & Peterson (1996)	259 firms (1988-1992)	Industrials (U.S.)			Compustat	Earnings ($r_S = 17\%$)
Bacidore et al. (1997)	600 firms (1982-1992)	Industrials (U.S.)	Yes	Yes	Stern Stewart	REVA ($R^2 = 4\%$)
Biddle et al. (1997)	773 firms (1983-1994)	Industrials (U.S.)	Yes	Yes	Stern Stewart	Earnings ($R^2 = 9\%$)
Chen & Dodd (1997)	566 firms (1983-1992)	Industrials (U.S.)			Stern Stewart	N/A
Clinton & Chen (1998)	325 firms (1991-1995)	Industrials (U.S.)			Stern Stewart	Cash flow ($r_S = 31\%$)
Günther et al. (2000)	36 firms (1989-1996)	All sectors (Germany)			BCG	Earnings ($r_S = 50\%$)
Turvey et al. (2000)	17 firms (1994-1996)	Food (Canada)		Yes	Market Watch Canada	N/A
Chen & Dodd (2001)	1,000 firms (1983-1992)	Industrials (U.S.)	Yes	Yes	Stern Stewart	Oper. income ($R^2 = 6\%$)
Fernández (2001)	100 firms (1994-1998)	All sectors (Global)			BCG	N/A
Pälchen & Schremper (2001)	362 firms (1989-1998)	S&P 400 (U.S.)		Yes	BCG, Data-stream	Δ CVA ($R^2 = 38\%$)
Sandoval (2001)	62 firms (1994-1999)	Industrials (Chile)		Yes	Economatica	EVA ($R^2 = 24\%$)
Stelter et al. (2001)	304 firms (1990-1995)	S&P 400 (U.S.)			BCG	Δ CVA ($r_S = 57\%$)
Fernández (2002)	28 firms (1992-1998)	Industrials (Spain)			Stern Stewart	EVA ($r_S = 28\%$)
Bernstein & Subramanian (2003)	500 firms (1986-2002)	S&P 500 (U.S.)			Merrill Lynch, Compustat	N/A
Feltham et al. (2004)	1,000 firms (1995-1999)	Industrials (U.S.)	Yes	Yes	Stern Stewart	RI ($R^2 = 5\%$)
Feltham et al. (2004)	300 firms (1991-1998)	Industrials (Canada)	Yes	Yes	Stern Stewart	EVA ($R^2 = 11\%$)
Copeland et al. (2004)	500 firms (1992-1998)	S&P 500 (U.S.)		Yes	Stern Stewart, BARRA	EPS ($R^2 = 6\%$)
West & Worthington (2004)	110 firms (1992-1998)	Industrials (Australia)			Stern Stewart, AGSM	EVA ($R^2 = 26\%$)
Anastassis & Kyriazis (2007)	107 firms (1996-2003)	All sectors (Greece)	Yes	Yes	Effect Finance, Datastream	Earnings ($R^2 = 17\%$)
Erasmus (2008)	386 firms (1991-2005)	Industrials (South Africa)	Yes	Yes	McGregor BFA	Earnings ($R^2 = 8\%$)

Note: Considerably differing levels of R^2 result from the individual model specifications.
* Firm performance and market value consistently scaled by book value.
** Regression model based on single firm-years and consistently applied.

Table 13   Studies Relating Performance Measures to Stock Returns

Other studies apply a more advanced regression analysis on data from U.S. firms, differentiating between relative and incremental information content.[330] I.e., relative information content analysis allows a ranking of performance measures for exclusive choices among performance measures and incremental information content indicates the value-relevance of component measures or supplemental disclosures.

In 1997, Biddle et al. concluded that relative and incremental information content of earnings is significantly greater than that of other measures: [331] Earnings explain returns significantly better than RI, EVA and operating cash flow (respective R^2s of 9%, 6%, 5% and 2%). After decomposing EVA into unique information components, the study shows earnings components provide highly significant and much greater incremental contributions to the explanation of stock returns than RI and EVA components. Several sensitivity analyses reconfirm the dominance of earnings.[332] Therefore, marginal incremental information content of value-based measures is assumed to be too low to provide greater relative information content.

In 2001, Chen and Dodd reconfirmed that earnings have greatest relative and incremental information content: [333] Relatively, operating income significantly outperforms RI and EVA, respective R^2s of 6%, 5% and 2%.[334] Incrementally, RI and EVA components increase R^2 by 3% and by just 1%, respectively. However, RI and EVA both have significant incremental information content at the 1% level. Slightly differing results compared to the study of Biddle et al. may refer to the following: The study applies differing methods to exclude extreme values, uses the 'level and changes' rather than the 'one-lag' model to estimate unexpected performance, omits the White correction for heteroscedasticity, and examines slightly differing time windows.

In 2004, Feltham et al. replicated the Biddle et al. study, repeating the analysis for 'surviving' firms and a more current but shorter period: [335] With 79 firms dropping out due to business combinations, business cessation, or poor performance, analysis based on observations from 694 'survivors' shows that value-based measures, EVA and RI, have greater relative information content than earnings or operating cash flow

---

[330] See Biddle et al. (1997) and Biddle, Seow, and Siegel (1995).

[331] See Biddle et al. (1997).

[332] After separating the independent variable into positive and negative values, the derived ranking remains, although the R^2 of earnings, residual income, EVA and operating cash flow of 13%, 7%, 6% and 3% increased for each measure. Extension of the performance and return period to a 5-year interval increases the association of variables, with R^2s of earnings, CFO, EVA and residual income of 31%, 19%, 14% and 11%, respectively. A sensitivity analysis that differentiates firms that adopted EVA does not include significant statements on relative information content and does not alter statements on incremental information content.

[333] See Chen and Dodd (2001).

[334] Superiority of operating income endures after applying yearly rather than pooled data, respective R^2s of 9%, 8% and 7%.

[335] See Feltham et al. (2004); there is a slight divergence from the Biddle et al. design, since Feltham et al. use respective period and not constant scaling factors for the contemporaneous and lagged performance.

(respective R^2s of 6%, 6%, 3% and 3%). Moreover, EVA significantly outperforms traditional measures on a 5% significance level. However, EVA components are not significant at the 5% level and results are generally affected by survivorship bias. Second, results based on a more current period show that RI and EVA dominate earnings and operating cash flow (respective R^2s of 5%, 4%, 3% and 2%). However, only RI has significant relative information content advantage versus operating cash flows at a 5% significance level, and the sample includes a considerably lower number of observations and covers only half the number of years as before.

In 2008, Erasmus replicated the Biddle et al. study for South African firms, additionally examining information content of depreciation-adjusted and inflation-adjusted metrics:[336] In line with the results of Biddle et al., information content of earnings ($R^2 = 8\%$) is clearly greater than information content of CFROI and RI (respective R^2s of 4%), operating cash flow and EVA (respective R^2s of 3%), and inflation-adjusted EVA and inflation-adjusted CVA (respective R^2s of 2%). While the CFROI and CVA metrics are independently computed, the EVA metric is obtained from the McGregor BFA database that does not include all accounting adjustments suggested by Stern Stewart.[337] The study is based on 198 listed and 188 delisted firms to reduce survivorship bias.

To summarize, more advanced studies on value-relevance provide evidence for earnings dominating EVA.

Further studies examine less common value-based metrics such as the modified EVA metric REVA, cash flow metrics TBR, CVA, and CFROI, and an expectations-based metric for U.S. firms.

In 1997, Bacidore et al. find that a newly-defined version of EVA, called 'Refined EVA' (REVA) is more value-relevant than the basic EVA metric, relating abnormal returns to scaled contemporaneous and lagged performance:[338] REVA (4%) explains two times the share of abnormal returns explained by EVA (2%), while R^2s are relatively low. Contrary to EVA, REVA implies capital expenditures based on the market value instead of book value of the firm's assets.[339] However, it is unclear whether the superiority of REVA stems from the modified capital based or the beta calculation, since REVA estimates beta differently than EVA.[340] However, Sandoval refutes superiority of REVA for Chilean firms, relating current and lagged performance to abnormal returns:[341] EVA shows the highest level of R^2 (24%), followed by REVA, net income, and operating cash flow, with respective R^2s of 23%,

---

[336] See Erasmus (2008).

[337] E.g., the estimated CVA measure equals NOPAT, depreciation, amortization and changes in other long-term liabilities less a capital charge on invested capital and accumulated depreciation. Consequently, it neither fully matches the approach described by Madden (see section 2.2.2) nor corresponds to one of the newly introduced measures in the study (see section 3.3.1).

[338] See Bacidore et al. (1997).

[339] Cost of capital calculation follows best practice (see Dimson, 1979; Fama & French, 1992).

[340] Cost of capital of EVA is based on individual firm performance, see Stewart (1991, p. 438). Cost of capital of REVA is derived from firm betas based on ten size-based portfolios.

[341] See Sandoval (2001).

3%, and 2%.[342] However, overall results are hardly reliable, since the underlying sample of 62 firms refers to an immature market and is rather small.

In 1996, Olsen finds that TBR is superior to EVA:[343] Changes in EVA are negatively and poorly correlated with stock returns by -0.16%. In fact, 68% of the firms had negative EVAs in 1994. Instead, TBR shows correlations with stock returns on one-year and three-year periods of 40% and 57%, respectively.

In 2001, Stelter et al. conclude that CVA is superior to EVA, based on 5-year data:[344] The change in EVA p.a. (33%) has lower correlation with TBR than change in CVA p.a. (57%).[345] However, high correlation results may result from using average rather than annual data. Schremper and Pälchen reconfirm the superiority of CVA, applying 10-year averages:[346] Changes in CVA divided by gross capital explain 38% of gross returns, while changes in EVA divided by net capital explain 22%. Contrarily, Fernández shows that value-relevance of CVA is fairly low, based on the 1994-1998 change in CVA and stock returns:[347] The correlation between changes in CVA and returns is 1.7%.

In 2003, Bernstein and Subramanian conclude that CFROI is not outperforming traditional metric:[348] While the S&P 500 index increased by 10.4%, top 50 CFROI firms (15%) outperformed by 4.6% and bottom 50 CFROI firms underperformed slightly by 0.7%. However, outperformance of the top 50 CFROI portfolio only prevails over time, exhibiting irregular levels of over- and underperformance. Furthermore, risk/return characteristics of alternative stock selection strategies based on earnings or return measures are not minor to those from CFROI portfolios.

In 2004, Copeland et al. showed that an expectations-based measure is superior to EVA and earnings:[349] Earnings-per-share and EVA, both scaled by the beginning market value, explain 6% and 2% of abnormal returns, respectively.[350] However, earnings expectations allow explaining 47% of abnormal returns.

To summarize, studies on alternate metrics to EVA provide some indications that REVA, TBR, CVA, and an expectations-based metric outperform EVA, while not being sufficiently robust.

---

[342] Moreover, results show that REVA explains returns best when applied to metallurgy (32%) and beverage (31%) firms, while pension funds, electric firms, investment firms, and construction firms show R^2s of 26%, 22%, 19%, and 8%, respectively.

[343] See Olsen (1996).

[344] See Stelter et al. (2001).

[345] Extending the time window to 10 years and decreasing the number of firms, correlation of changes in EVA and CVA p.a. further increases to 50% and 68%, respectively.

[346] See Schremper and Pälchen (2001).

[347] See Fernández (2001).

[348] See Bernstein and Subramanian (2003).

[349] See Copeland et al. (2004).

[350] Equivalently, growth in earnings-per-share and growth in EVA explain 5% and 4% of abnormal returns, respectively. After a logarithmic transformation of all independent variables, earnings still has higher value relevance than EVA.

Finally, value-relevance of value-based performance measures has also been examined for non-U.S. firms.

In 1999, Günther et al. reconfirm superiority of earnings for 36 German firms:[351] Traditional performance measures show highest correlations, where return on equity, return on sales, return on investments, CVA and EVA show correlation coefficients of 50%, 47%, 35%, 31%, and 2%, respectively, with stock returns.

In 2000, Turvey et al. examine value-relevance of EVA for 17 Canadian food processing firms: [352] Results show that EVA has a relatively low adjusted R^2 of 7%. However, restrictive sector and geographic limitations, a small sample, and omission of competing performance measures relativize their results.

In 2002, Fernández demonstrates outperformance of EVA for the 28 largest Spanish firms: [353] EVA (28%) is more highly correlated with returns than after-tax profits (19%) and return on equity (11%).

In 2004, Feltham et al. reproduce the study of Biddle et al. for 300 Canadian firms, finding further evidence for the dominance of EVA:[354] At a 5% significance level, EVA is superior to earnings, operating cash flow and residual income (respective R^2s of 11%, 3%, 1% and 0%). However, results based on solely 386 observations over a 7-year period have a purely indicative character.

Moreover, West and Worthington show that EVA stands out explaining returns for 110 Australian firms, based on a fixed-effects model: [355] EVA, RI, earnings and operating cash flows explain 26%, 19%, 14% and 14% of returns, respectively. However, the study does not include statements on significance. Additionally, the fixed-effects model imposes limitations on interpreting results.

In 2007, Anastassis and Kyriazis demonstrate the dominance of earnings for 107 Greek firms, using a market model to estimate excess returns:[356] Based on pooled data, operating income, net earnings, residual income and EVA explain 17%, 9%, 8% and 7% of abnormal returns, respectively. [357] Furthermore, differences in relative

---

[351] See Günther, Landrock, and Muche (1999).

[352] See Turvey, Lake, van Duren, and Sparling (2000).

[353] See Fernández (2002a).

[354] See Feltham et al. (2004).

[355] See West & Worthington (2004). The study stands out by providing a profound test of three pooling models for panel data. Extensively applied in practice, the common-effects model assumes that financial relations are homogeneous across firms. Alternatively, a fixed-effects model allows a different intercept by observation and a random-effects model assumes coefficients are random variables. Following the fixed-effects model, association of all variables with returns is 28%, 19% based on the random-effects model and 3% under the common-effects model. Thus, the low explanatory power of the common-effects model may point to an incorrect model specification applied within other studies.

[356] See Anastassis and Kyriazis (2007).

[357] Results prevail when examining the association of performance with raw returns (respective R^2s of 17%, 8%, 8% and 6%). Average results from annual regressions reveal lower associations for income measures and, consequently, lower absolute differences in associations of competing measures: Operating income, net earnings, RI and EVA explain 8%, 6%, 6% and 5% of abnormal returns and 10%, 7%, 7% and 7% of raw returns, respectively.

information content are all significant at least at a 5% level except the greater association of RI than EVA with abnormal returns. Finally, unique earnings and RI components add significant incremental information content at a 1% level; incremental EVA components do not add significant information advantage.

All in all, results for non-U.S. firms are ambiguous, primarily derived from insufficient observations or, in one case, from a non-standard regression model.

To summarize, return studies for the most part provide clear evidence for earnings outperforming EVA and also some indications that CVA is superior to EVA. While Feltham et al. provide some contrary evidence, reference studies from Bacidore et al., Biddle et al., and Chen and Dodd, going back to a comparable statistical approach, consistently show that earnings (R^2s ranging from 6% to 9%) dominate EVA (R^2s ranging from 2% to 5%). Furthermore, outperformance of earnings persists after controlling for the stage of the business cycle, i.e. in expansion versus contraction periods, and is more evident for manufacturing than non-manufacturing firms.[358]

4.1.3   Gaps in Previous Studies

The previous consolidation of available value-based performance measures and corresponding value-relevance studies reveals several deficiencies in the definition of metrics and the conception of studies, as follows:

Derived from the VBM objective to maximize shareholder value and, thus, reconciling with the firm's market value, value-based performance measures assist management and the financial community in quantifying economic value creation. However, common single-period VBM measures of firm performance either lack a hurdle rate, or suffer from age bias, or entail considerable complexity, specifically:

Metrics such as CFROI, ROCE, or ROIC, describe overall firm performance, while omitting to integrate a threshold margin to determine whether a firm added or destroyed shareholder value. Metrics such as RI and EVA include a threshold margin but show a relative performance increase over the assets' life, thus, being distorted by age bias.[359] Representing the only common measure expressing shareholder value creation and unaffected by the age of the assets, CVA optimally reflects shareholder value creation, while introducing substantial complexity that impedes its successful implementation.

Apart from theoretical foundations, the business community is primarily concerned about the statistical validity of performance measures to explain stock prices. While there was some consistent evidence on the outperformance of earnings, latest evidence on EVA challenges previous results. Further, empirical studies commonly lack regression statistics, apply inconsistent regression models, use publicly unavailable

---

[358] See results from Kim, Ahn and Yun (2004, Spring) based on a sample of 1,109 firms included in the Stern Stewart ranking database for the years 1990 to 1995.

[359] Consecutively, I refer to the EVA measure rather than the EP measure, since both represent RI measures adjusted for accounting distortions, being almost identically defined, while EVA is predominantly used.

databases, omit significance tests, or do not integrate results from relative and incremental information content and associations with market value and stock returns, as follows:

Frequently, value-relevance studies provide solely correlation statistics that do not allow inferences on causality. [360] While regression analysis allows inferences on causality, e.g., O'Byrne and Tsuji inconsistently test EVA based on one regression model while other measures than EVA on a differing regression model, imposing difficulties to compare regression results from competing measures. Further, studies with an extensive underlying sample commonly use in-house data provided by consultancies, specifically Stern Stewart or Boston Consulting Group data, although inherent computation principles are unknown.[361] Apart from underlying data, studies commonly provide only levels of R^2 to describe relative information content while omitting significance tests on differences in information content from competing measures. [362] Complementing tests on incremental information content provide additional insights; however, these tests are commonly omitted. Finally, literature examines either associations with market value or associations with stock returns, however, rarely combining these two approaches.

Consequently, deficiencies summarized above propose to extend research twofold:

• First, the definitional contribution of this study comprises to adjust RI and EVA, to avoid inherent age bias.

• Second, the empirical contribution of this study involves reexamining the value-relevance of prevalent and recommended performance measures using a profound study design and enhanced data, to overcome limitations of previous research.

Figure 21 illustrates the contributions of the present study, outlining characteristics of eight reference studies.

---

[360] Of 27 reviewed studies on competing performance measures, 8 (30%) limit statistics to correlation coefficients.
[361] Of 24 reviewed studies that examine data from at least 100 firms, 16 (67%) use Stern Stewart ranking data, four studies use the BCG database, one study applies Merrill Lynch data, and, finally, one correlation analysis on U.S. firms and two regression analyses on Japanese and Greek firms use independently computed data.
[362] While statistical standard software provides p-values on incremental information content, it requires additional calculations to derive p-values on relative information content

	Outperforming metric according to R² lead	Usable VBM metric excluding age bias	Replicable data	Consistent regression model	Relative and incremental analysis	Significance tests on differences in R²	Associations with value and returns
Present study	N/A	+	+	+	+	+	+
O'Byrne (1996)	EVA	-	-	-	-	-	-
Biddle et al. (1997)	Earnings	-	-	+	+	+	+
Kramer & Pushner (1997)	Earnings	-	-	+	-	-	-
Chen & Dodd (2001)	Earnings	-	-	+	-	-	-
Pälchen & Schremper (2001)	CVA	-	-	+	-	-	-
Feltham et al. (2004)	RI EVA	-	-	+	+	+	-
Copeland et al. (2004)	Earnings	-	-	+	-	-	-
Tsuji (2006)	Cash flow	-	+	-	-	-	-

Note: Benchmark studies selected requiring a regression analysis on a minimum of 150 firms, comparing the value-relevance of at least one traditional and value-based performance measure.

Figure 21 Characteristics of Reference Studies Compared to the Present Study

4.2 Development of the Empirical Study Design

4.2.1 Research Hypotheses

Having introduced CRI and CEVA, this study examines the value-relevance of earnings, operating cash flows, RI, CRI, EVA, and CEVA. Statistical analysis of value-relevance assumes that capital markets are (semi-strong) efficient and sophisticated, i.e., forming future estimates of performance measures. Starting point are four research questions, which require examining value-relevance on the basis of either market values or stock returns and analyzing value-relevance from either a relative or incremental view.[363]

The first relative information content question examines whether VBM measures of firm performance, RI, CRI, EVA, and/or CEVA, dominate currently mandated performance measures, earnings and operating cash flows, in explaining average annual market values of the firm, as addressed by the following neutral null hypothesis:

Null hypothesis $H_0^{(1)}$: Examining associations with firm value, the information content of performance measure X_1 is indeed superior to that of X_2.

Corresponding analysis will provide a direct test of latest evidence on the outperformance of EVA,[364] and quantify value-relevance of newly introduced measures CRI and CEVA. Theoretical considerations support the superiority of CEVA: Foremost, valuation theory suggests that residual income measures explain firm value better than earnings or operating cash flows, both not supported by financial theory.[365] Moreover, EVA and CEVA, minimizing distortions from a number of accounting items, should outperform RI and CRI in explaining market values that are assumed to reflect the long-term ability to create value. CEVA, reducing additionally distortions from accounting depreciation, should be superior to EVA.

---

[363] Researchers commonly employ a combination of relative and incremental information content tests, having differing practical implications (see Biddle et al., 1997; Biddle et al., 1995; R. M. Bowen et al., 1987; Chen & Dodd, 2001; G. P. Wilson, 1986, 1987).

[364] Some of the latest value-relevance studies suggest that, more recently, markets reflect RI and/or EVA rather than earnings, see section 4.1.

[365] See section 2.2.2 and 4.2.2.

The second research question, referring to incremental information content is whether financial markets value unique information components of earnings, RI, CRI, EVA, and/or CEVA beyond the information contained in cash flows. Consequently, CRI and CEVA are decomposed into unique information components (see section 4.2.3) and the contribution of each component is evaluated towards explaining contemporaneous annual firm values, based on the following null hypothesis:

Null hypothesis $H_0^{(2)}$: Component C_i does not provide information content beyond that provided by the remaining components C_1–C_{i-1} in explaining variations of firm value.

Again relating to relative information content, the third research question is whether VBM measures, RI, CRI, EVA, and/or CEVA, outperform currently mandated performance measures, earnings and operating cash flows, in explaining contemporaneous annual stock returns. Even though market participants may continue to favor earnings, again, a neutral null hypothesis is used:

Hypothesis $H_0^{(3)}$: Examining associations with stock returns, the information content of performance measure X_1 is indeed superior to that of X_2.

This hypothesis, being based on stock returns, complements the first research hypothesis, being based on market value, providing an alternative quantification of value-relevance. Practice suggests superiority of earnings in explaining stock returns due to earnings myopia of financial markets: Commonly, investors behave myopic, sacrificing long-term wealth creation for short-term earnings improvements.

The fourth empirical question is whether components unique to earnings, RI, CRI, EVA, and/or CEVA, aid explaining contemporaneous annual stock returns beyond that portion yet explained by operating cash flows. Again, components of CRI and CEVA are used to assess incremental information content, based on the following null hypothesis:

Hypothesis $H_0^{(4)}$: Component C_i does not provide information content beyond that provided by the remaining components C_1–C_{i-1} in explaining variations of stock returns.

To test value-relevance based on these four null hypotheses, two regression equations as well as two statistical tests are applied, as introduced in the following sections.

4.2.2   Regression Models Relating Performance to Market Value and Returns

The following empirical study applies two regression equations that relate accounting-based value performance to either the level of the stock price (more precisely, the market value of the firm) or the change in the stock price (i.e., stock returns).

Model Relating Performance to Market Value

The underlying relationship of the first regression model, which relates accounting-based firm performance to the market value of the firm, is derived from traditional valuation theory.

The 'Discounted Cash Flow' (DCF) model and the Dividend Discount model, both assuming efficient capital markets, are the most common approaches to corporate valuation. According to the DCF model, the end-of-year corporate value of total (debt and equity) capital, CV, is equal to the sum of expectations, E, of future free cash flows returned to all investors (debtholders and shareholders), FCF, discounted by an appropriate risk-adjusted rate, r, as follows:

$$CV_t = \sum_{j=1}^{\infty} \frac{E_t(FCF_{t+j})}{(1+r)^j} \cdot {}^{366}$$

In contrast to the DCF model, the Dividend Discount model neglects the value of debt capital and applies dividends as a measure of future cash flows expected to be available to shareholders. It defines the end-of-year corporate value of an all-equity firm (or stock value, P) as sum of infinitive future dividends, D, discounted by an appropriate risk-adjusted rate (the cost of equity, r_e), notationally:

$$P_t = \sum_{j=1}^{\infty} \frac{E_t(D_{t+j})}{(1+r_e)^j} \cdot {}^{367}$$

However, neither free cash flows nor dividends describe the periodic value creation of the firm as demanded by the VBM concept. Free cash flows disregard the cost of debt and equity capital in order to measure value added to shareholders and, therefore, implicitly measures the value creation with respect to a zero capital level. Dividends reflect the management decision to return free cash to shareholders rather than to invest in additional projects or reinvest in existing operations, so that dividends may be generally affected by investment policies also related to growth opportunities.

---

[366] See Brealey and Myers (2000, pp. 77-78). The term corporate value implies that actual market values may differ. The formula assumes a flat term-structure of discount rates.
[367] See Brealey and Myers (2000, p. 66).

'Residual Income' (RI), defined as $RI_t = NE_t^{BI} - WACC_t \cdot NA_{t-1}$, seems a more valid measure of economic value creation in a VBM context.[368] According to the Edwards-Bell-Ohlson model (also referred to as residual income valuation relationship), the end-of-year market value of a stock can be rewritten to equal the end-of-year book value of the firm, BV, and an infinite sum of discounted future residual incomes, RI, notationally:

$$P_t = BV_t + \sum_{j=1}^{\infty} \frac{E_t(RI_{t+j})}{(1+r_e)^j}.[369]$$

The RI model assumes that the clean surplus relation holds, which states that the change in book value, BV, from period to period equals net earnings minus net dividends, notationally $BV_t = BV_{t-1} + NE_t - D_t$.[370] In practice, the clean surplus assumption is quite reasonable.[371]

Stewart assumes that the RI valuation relationship still holds for EVA, notwithstanding that EVA contains a series of accounting adjustments.[372] In fact, the DCF and RI valuation relationships introduced above have been extended to encompass a number of modified residual income measures such as EVA, CVA and Economic Profit and several cash flow measures such as equity cash flows, capital cash flows, risk-adjusted cash flows, or risk-free-rate-adjusted cash flows.[373] Further, the DCF and RI models reconcile with and without taxes,[374] and under varying leverage or terminal value assumptions.[375] Finally, researchers provided empirical evidence on the association of RI with market value.[376]

---

[368] See section 2.2.2.

[369] Several authors discuss the residual income valuation relationship (see Bernard, 1994; Edwards & Bell, 1961; Ohlson, 1990, 1995; Preinreich, 1938).

[370] Ohlson (1995) proved that accounting systems are required to fulfill the clean surplus relation to tie to the discounted residual income model. Then, the Edwards-Bell-Ohlson model arises from rewriting the Dividend Discount model by substituting dividends by residual income via the clean surplus relation, as follows:

$$CV_t = \sum_{j=1}^{\infty} \frac{D_{t+j}}{(1+r_e)^j} = \sum_{j=1}^{\infty} \frac{NE_{t+j} + BV_{t+j-1} - BV_{t+j}}{(1+r_e)^j} = \sum_{j=1}^{\infty} \frac{RI_{t+j} + (1+r_e) \cdot BV_{t+j-1} - BV_{t+j}}{(1+r_e)^j} = BV_t + \sum_{j=1}^{\infty} \frac{RI_{t+j}}{(1+r_e)^j}.$$

[371] Feltham and Ohlson (1995), Frankel and Lee (1998) and Krotter (2007) provide evidence that 'dirty surplus' items, distorting the clean surplus relation, have no major impact on the relation in practice.

[372] See Stewart (1991, pp. 153-154) and section 2.3.3. Specifically, Stewart explains firm value (instead of stock value) via the invested capital, BV, and the expected future EVAs discounted by the weighted average cost of debt and equity capital, WACC, (instead of the equity cost), notationally: $CV_t = BV_t + \sum_{j=1}^{\infty} \frac{EVA_{t+j}}{(1+WACC)^j}$.

[373] See Fernández (Fernández, 2002b), and Shrieves and Wachowicz (2001).

[374] See Tham (2001a).

[375] See Tham (2004).

[376] See Biddle, Chen, and Zhang (2001), Frankel and Lee (1998), Lee, Myers, and Swaminathan (Lee, Myers, & Swaminathan, 1999), and Penman and Sougiannis (1998).

Derived from the DCF and RI model described above, the following equation defines the univariate linear regression model used to evaluate value-relevance of any performance measure, X (avoiding different model specifications for different performance measures), where MV = market value of the firm, BV = book value of total assets, and $WACC$ = weighted average cost of capital:

$$\frac{MV_t}{BV_{t-1}} = \beta_0 + \beta_1 \cdot \frac{X_t / WACC_t}{BV_{t-1}} + \varepsilon_t.$$

This model is consistently applied across performance measures and has two distinct characteristics:[377]

- First, it applies the perpetuity of the current performance instead of individual estimates of future performances. Since major databases (such as Value Line) do not include expectations on RI or residual income variants such as EVA, performance is assumed to be constant, to avoid subjective estimates of future performance.

- Second, it divides both sides of the equation by beginning book value of net capital, omitting the book capital as second independent variable.[378] Then, percentage errors are more likely to be closer to a normal distribution than prediction errors in absolute terms. Furthermore, a scaled model can be interpreted as DCF or RI model: an estimated intercept of zero would reconcile it with the DCF model; an estimated intercept of one would match it to the RI model.

In addition to the basic regression specification introduced above, four model extensions serve to analyze the sensitivity of results, as follows:

- Assuming a slope coefficient of one, the basic model specification above implies that current-year performance endures infinitively. However, firms with positive residual income are likely to increase their capital base if their earnings exceed the capital expenditures. Consequently, the slope coefficients are most likely greater

---

[377] Previous studies commonly applied one model consistently across metrics, see Biddle et al. (1997) and Kramer and Pushner (1997). By contrast, O'Byrne (1996, p. 120) requires a zero intercept for testing value-relevance of earnings, arguing that a non-zero intercept allows an earnings model to convert into a RI model, as follows. $\frac{MV_t}{BV_{t-1}} = a + b \cdot \frac{NE_t}{BV_{t-1}}$ is rewritten as

$MV_t = a \cdot BV_{t,t-1} + b \cdot NE_{t,t} = a \cdot BV_{t-1} + b \cdot NE_t + (1-a) \cdot BV_{t-1} - (1-a) \cdot BV_{t-1} =$

$= BV_{t-1} + b \cdot NE_t - (1-a) \cdot BV_{t-1} = BV_{t-1} + b \left[NE_t - BV_{t-1} \cdot \frac{(1-a)}{b} \right] = BV_{t-1} + b \cdot RI$,

where RI' denotes a residual income measure with a cost of capital of $(1-a)/b$; it follows $\frac{MV_t}{BV_{t-1}} = 1 + b \cdot \frac{RI'_t}{BV_{t-1}}$. However, resemblance with the Edwards-Bell-Ohlson model does not seem sufficient to require a zero intercept. Reconciliation following O'Byrne of the earnings and RI model requires that the cost of capital equals $(1-a)/b$. Further, the Edwards-Bell-Ohlson model relates the perpetuity of RI to market value, while O'Byrne omits the capitalization of RI. Finally, the earnings and RI model introduced by O'Byrne differ by the expected intercept.

[378] Researchers commonly scale dependent and independent variables   (see Kramer & Pushner, 1997; O'Byrne, 1996; G. B. Stewart, 1991).

than one. On the other hand, firms with negative residual income are unlikely to continue additional investments that do not earn the capital expenditures. In addition, investors' expectations on future turnarounds may involve a slope coefficient less than one, or even negative. Additionally, studies provide evidence that loss firms demonstrate smaller slope coefficients.[379] Consequently, the basic regression equation is modified to account for differing investor expectations based on the sign of performance, as follows:[380]

$$\frac{MV_t}{BV_{t-1}} = \beta_0 + \beta_1 \cdot \frac{X_t^{pos} / WACC_t}{BV_{t-1}} + \beta_2 \cdot \frac{X_t^{neg} / WACC_t}{BV_{t-1}} + \varepsilon_t.$$

- While capitalization of performance by the cost of capital is necessary to match up with valuation theory, it involves major errors of estimates. Therefore, the model shall be retested without capitalizing performance, in order to reduce errors from perpetuity assumptions, as follows:[381]

$$\frac{MV_t}{BV_{t-1}} = \beta_0 + \beta_1 \cdot \frac{X_t}{BV_{t-1}} + \varepsilon_t.$$

- Since listed U.S. firms take up to three months until filing annual financial statements, it may be necessary to allow more time for stock prices to impound accounting information.[382] Therefore, the basic model is also adjusted to incorporate average market value over the contemporaneous and the subsequent fiscal year instead of current-year market value, as follows:

$$\left(\frac{MV_t}{BV_{t-1}} + \frac{MV_{t+1}}{BV_t} \right) / 2 = \beta_0 + \beta_1 \cdot \frac{X_t / WACC_t}{BV_{t-1}} + \varepsilon_t.$$

- Finally, G. Bennett Stewart, founding member of Stern Stewart & Company, claimed that EVA shows the maximum information content on five-year data.[383] To address Stewart's claim, the basic model is modified to incorporate contemporaneous 5-year changes of performance and firm value, as follows:

$$\frac{MV_{t+5}}{BV_{t+4}} - \frac{MV_t}{BV_{t-1}} = \beta_0 + \beta_1 \cdot \left(\frac{X_{t+5} / WACC_{t+5}}{BV_{t+4}} - \frac{X_t / WACC_t}{BV_{t-1}} \right) + \varepsilon_t.$$

---

[379] Evidence prevails for earnings (see Burgstahler & Dichev, 1997; Collins, Pincus, & Xie, 1999; Hayn, 1995) and RI (see Giner & Iniguez, 2006).

[380] A modification to account for sign of performance is commonly used (see Biddle et al., 1997; Hayn, 1995; O'Byrne, 1996).

[381] Research commonly applies such a regression model (see Biddle et al., 1997; Kramer & Pushner, 1997; Tsuji, 2006).

[382] See Securities and Exchange Commission. RIN 3235-AI33: Acceleration of Periodic Report Filing Dates and Disclosure Concerning Website Access to Reports; Securities and Exchange Commission. RIN 3235-AJ29: Revisions to Accelerated Filer Definition and Accelerated Deadlines for Filing Periodic Reports.

[383] See Stewart (1991, p. 742) and Stewart (1994).

Model Relating Performance to Returns

The second regression model relates accounting performance to stock returns. Specifically, the scaled performance level and change shall explain abnormal returns, AR, based on an accounting-based performance measure, X, and the market value of the firm's equity, MVE, as follows:

$$AR_t = \beta_0 + \beta_1 \cdot \frac{X_t}{MVE_{t-1}} + \beta_2 \cdot \frac{(X_t - X_{t-1})}{MVE_{t-1}} + \varepsilon_t.$$

The model specified above has three distinctive characteristics:

- First, it uses abnormal stock returns (beyond market-wide returns) instead of unadjusted stock returns of the firm as dependent variable. Systematic market returns shall be excluded to avoid spurious correlation with firm performance due to large correlated market components and to allow comparison over time and across firms. Adjustment techniques to eliminate systematic market returns include residuals from an estimated market model,[384] size-adjusted returns,[385] and market-adjusted returns.[386] In this study, arithmetic market-adjusted returns shall be applied, representing excess returns from investing in the common stock of the firm compared to an investment in the market index.[387]

- Second, it applies the 'one-lag' model, using the stock price level and the price change versus the prior year, to approximate the unexpected performance. Elimination of performance expectations yet impounded in the initial stock prices fulfils with equity valuation theory (suggesting unexpected deviation from performance expectations, i.e., unexpected performance, determines stock returns) and reduces heteroscedasticity over time, related to growth and inflation. Analysts' forecasts,[388] the 'one-lag' model,[389] and the 'level and changes' model[390] are

---

[384] Residuals are based on a regression of firm returns with returns from an equally-weighted or value-weighted market-wide index, see Bowen et al. (1987), Easton and Harris (1991), Patell (1976), Peterson and Peterson (1996), Rayburn (1986), and Strong and Walker (1993, April)

[385] Size-adjusted returns equal total stock returns less equally weighted average returns of size deciles based on the market value of the equity, see Peterson and Peterson (1996).

[386] Market-adjusted returns are excess returns excluding value-weighted market-wide index returns. Commonly, arithmetic market-adjusted returns are measured as 12-month compounded stock return less 12-month compounded value-weighted market-wide return, see Biddle et al. (1995, 1997), Bowen, Johnson, and Shevlin (1989), and Feltham et al. (2004). However, Fama argues that arithmetic abnormal returns are distorted over a longer time period, see Fama (1998). E.g., an arithmetic abnormal return of 20% from periodic security return of 10% and 200% ($1.1 \cdot 2 = 2.2$) and a cumulative index return of 200% overall is higher than the geometric abnormal return of 10% ($2.2/2.0-1=0.1$). Therefore, Copeland et al. suggest adopting geometric market-adjusted returns, equaling 12-month compounded stock return divided by 12-month compounded value-weighted market index returns, see Copeland et al. (2004).

[387] See section 4.2.5.

[388] E.g., Foster (1986) applies analyst forecast to eliminate market expectations.

[389] See Easton and Harris (1991). Independent from theoretical models, Easton and Harris find that the level and change of earnings is most highly associated with stock returns, as follows: $R_t = a_0 + a_1 \cdot X_t + a_2 \cdot (X_t - X_{t-1}) + \varepsilon_t$. While the change in performance estimates unexpected

commonly used to eliminate performance expectations. Notably, the 'level and changes' model and the 'one-lag' model can be easily reconciled.[391] In this study, the 'one-lag' model has been chosen, because it is expected to imply lower multicollinearity of the independent variables than the 'level and changes' model and is applicable across various accounting-based measures (quite contrary to analysts' forecasts that are mostly unavailable for measures other than earnings).

- Third, it scales accounting performance by the beginning market value of equity, to reduce variance of error that tends to increase with corporate size, to ensure consistency with the return variable that is measured in percentage terms, and to remove individual performance expectations impounded in the market value. Accounting-based performance is commonly scaled by the beginning market value of equity.[392] Alternatively, it can be scaled by the consumer price index.[393] While deflation by a price index aims to reduce market expectations approximated by the inflation factor in order to remove some bias from time-series data, it does not reflect individual firm characteristics.

---

earnings, the level of performance captures transitory components, thus, increasing the explanatory power of the model, see Christie (1987).

[390] Biddle et al. deduce the model mathematically, see Biddle et al. (1997, pp. 308-310): Starting point is the standard regression equation, in which the significance of the slope coefficient, b_1, measures information content: $R_t = b_0 + b_1 \cdot FE_{X,t} + \varepsilon_t$, where, R_t is the dependent return variable for a time period t, $FE_{X,t}$ is the unexpected realization or forecast error for a performance measures X and ε_t is a random disturbance term. Following, the forecast error is defined as difference between realized performance and expectation of the market, notationally $FE_t = X_t - E(X_t)$. Assuming a discrete linear stochastic process with δ as a constant and Φ's as autoregressive parameters, the equation above can be rewritten as follows: $R_t = b_0 + b_1 \cdot (X_t - E(X_t)) + \varepsilon_t = b_0 + b_1 \cdot (X_t - (\delta + \phi_1 \cdot X_{t-1} + \phi_2 \cdot X_{t-2} + \phi_3 \cdot X_{t-3} + ...)) + \varepsilon_t$, rewritten as $R_t = b'_0 + b'_1 \cdot X_t + b'_2 \cdot X_{t-1} + b'_3 \cdot X_{t-2} + ... + \varepsilon_t$, where $E(b'_0) = b_0 - b_1 \cdot \delta$, $E(b'_1) = b_1$ and $E(b'_i) = -b_i \Phi_{i-1}$ for $i > 1$. Consequently, the final equation estimates market expectations jointly via the slope coefficients b'_i with an indefinitive number of lagged observations. To reduce complexity, Biddle et al. suggest limiting the number of lagged observations to one, as follows: $R_t \approx b_0 + b_1 \cdot X_t + b_2 \cdot X_{t-1} + \varepsilon_t$.

[391] $a_0 + a_1 \cdot X_t + a_2 \cdot (X_t - X_{t-1}) = a_0 + a_1 \cdot X_t + a_2 X_t - a_2 X_{t-1} = a_0 + (a_1 + a_2) \cdot X_t - a_2 \cdot X_{t-1}$, where b $= a_1 + a_2$ and $b_2 = -a_2$. Following, the expected sign of the coefficients differs based on the specified model: 'Level and changes' coefficients, a_1 and a_2, are both supposed to be positive, since positive firm performance and an increase in performance shall relate to an increase in stock price. Consequently, the reconciliation above demonstrates that 'one-lag' coefficient b_1 shall be positive, while respective coefficient b_2 shall be negative.

[392] Authors commonly scale independent variables by market value of equity (Bacidore et al., 1997; Biddle et al., 1997; Chen & Dodd, 2001; Christie, 1987; Copeland et al., 2004; G. D. Feltham et al., 2004; Rayburn, 1986).

[393] See Lipe (1986).

Following the same rationales as before, the basic regression model shall be extended to incorporate sign-based coefficients, to integrate two-year returns, and to evaluate five-year data.

- First, the basic equation adapts to positive and negative accounting performance, as follows:

$$AR_t = \beta_0 + \beta_1 \cdot \frac{X_t^{pos}}{MVE_{t-1}} + \beta_2 \cdot \frac{(X_t - X_{t-1})^{pos}}{MVE_{t-1}} + \beta_3 \cdot \frac{X_t^{neg}}{MVE_{t-1}} + \beta_4 \cdot \frac{(X_t - X_{t-1})^{neg}}{MVE_{t-1}} + \varepsilon_t.$$

- Then, the return measurement period is lagging the fiscal year by 15 rather than three months, to allow time for stock prices to impound accounting information:

$$(1 + AR_t) \cdot (1 + AR_{t+1}) - 1 = \beta_0 + \beta_1 \cdot \frac{X_t}{MVE_{t-1}} + \beta_2 \cdot \frac{(X_t - X_{t-1})}{MVE_{t-1}} + \varepsilon_t.$$

- Moreover, the model is extended to examine five-year periods, as follows:

$$(1 + AR_t) \cdot (1 + AR_{t+1}) \cdot (1 + AR_{t+2}) \cdot$$

$$\cdot (1 + AR_{t+3}) \cdot (1 + AR_{t+4}) - 1 = \beta_0 + \beta_1 \cdot \left(\frac{X_t}{MVE_{t-5}} + \frac{X_{t+1}}{MVE_{t-4}} + \frac{X_{t+2}}{MVE_{t-3}} + \frac{X_{t+3}}{MVE_{t-2}} + \frac{X_{t+4}}{MVE_{t-1}} \right) +$$

$$+ \beta_2 \cdot \left(\frac{X_{t-4} - X_{t-5}}{MVE_{t-5}} + \frac{X_{t-3} - X_{t-4}}{MVE_{t-4}} + \frac{X_{t-2} - X_{t-3}}{MVE_{t-3}} + \frac{X_{t-1} - X_{t-2}}{MVE_{t-2}} + \frac{X_t - X_{t-1}}{MVE_{t-1}} \right) + \varepsilon_t.$$

- Finally, the application of abnormal returns aims to eliminate systematic market returns, while implying potential errors to incorrectly reflect market components of individual stock returns. Therefore, the regression model is revised to relate performance to raw returns, RR, (that do not include deductions for market returns) instead of abnormal returns, AR, circumventing errors from false return adjustments, as follows:

$$RR_t = \beta_0 + \beta_1 \cdot \frac{X_t}{MVE_{t-1}} + \beta_2 \cdot \frac{(X_t - X_{t-1})}{MVE_{t-1}} + \varepsilon_t.$$

4.2.3 Relative and Incremental Information Content Tests

Regression equations introduced in the section above are applied to examine associations of performance measures.[394] In this sense, testing association allows to estimate stock prices from accounting-based performance measures without restrictions on the intercepts or coefficients of the regression function, recognizing that the market may apply stock valuation techniques that encompass other components or mechanisms than those defined by financial theory. Tests for the accuracy of performance measures in explaining stock prices are beyond the scope of this study.[395] To test associations of performance measures, two statistical tests, namely relative und incremental information content tests, are applied.

Relative Information Content Test Providing Competitive Rankings

Relative tests compare the association of each measure with firm value or returns, providing a ranking and, thus, a mutually exclusive selection criterion for competing measures. Hereby, the absolute level of adjusted R^2 and the significance of the difference in adjusted R^2 determine the relative information content of each variable.

According to hypotheses H_1 and H_3, relative tests involve comparing the levels of adjusted R^2s from six separate regressions, one for each measure. R^2s are calculated, applying White's correction for heteroskedastic errors,[396] based on the following basic regression models (where X_i are CFO, EBEI, RI, CRI, EVA, and CEVA):

$$\frac{MV_t}{BV_{t-1}} = \beta_0 + \beta_1 \cdot \frac{X_{i,t} / WACC_t}{BV_{t-1}} + \varepsilon_t \quad \text{and}$$

$$AR_t = \beta_0 + \beta_1 \cdot \frac{X_{i,t}}{MVE_{t-1}} + \beta_2 \cdot \frac{(X_{i,t} - X_{i,t-1})}{MVE_{t-1}} + \varepsilon_t$$

Moreover, H_1 and H_3 imply one-tailed tests of the null hypotheses that competing measures have differing information content, where X_1 and X_2 are pairwise combinations from the set of accounting measures. Rejection of H_1 or H_3 provides evidence of a significant equality in relative information content. To assess significant information advantage, the Cox test for non-nested regressions examines the null hypothesis of a difference in the ability of two sets of independent variables to explain variation in the dependent variable.[397]

---

[394] See previous research discussed in section 4.1.1.

[395] See Holthausen and Watts (2001) for the classification of association versus accuracy tests. A measure is supposed to be highly accurate if it fulfils relationships set by valuation theory. Evaluating accurateness imposes restrictions on the parameters of the regression functions, requiring an estimated intercept of zero or one for free cash flows or residual income measures, respectively, and an estimated slope coefficient of one.

[396] See White (1980).

[397] See Cox (1961); Cox's test statistic, q01, comparing independent variables X and Z, is distributed as standard normal N (0,1) and defined based on four regressions, as follows:

The same approach is applied to examine the sensitivity of the basic model previously defined model extensions.

Incremental Information Content Test Evaluating Additive Components

While relative information content tests examine accounting measures per se, incremental information content tests evaluate 'additive' information components between different accounting measures. Using CFO as base, several increments reconcile CRI and CEVA with CFO, as shown in Figure 22, namely accruals (*Accr*), after-tax interest (*ATInt*), capital charges (*CapChg*), accounting adjustments (*AccAdj*) and depreciation adjustments (*DepAdj^1/ DepAdj^2*).

Incremental tests examine whether one measure adds to the information provided by another in explaining market value or returns, evaluating the benefit of supplemental disclosures in financial reporting. Whereas, the absolute change in adjusted R^2 and the significance of the additional information, as described by the p-value from two-tailed F-tests of statistical significance, determines the incremental information content of each component variable. Interpretation of the change in R^2 rather than coefficients follows common statistics principles.[398]

Following H_2 and H_4, incremental tests require evaluating the increase in adjusted R^2 from the adjusted R^2 of common information components to the adjusted R^2 of common components and the additional unique information components. Increases in R^2 of each of the six unique components, Ci, are computed using White's correction for heteroskedastic errors, based on the following equation:[399]

$$\frac{MV_t}{BV_{t-1}} = \beta_0 + \beta_1 \cdot \frac{X_t^{C1}/WACC_t}{BV_{t-1}} + ... + \beta_1 \cdot \frac{X_t^{Ci}/WACC_t}{BV_{t-1}} + \varepsilon_t \text{ and}$$

$$AR_t = \beta_0 + \beta_1 \cdot \frac{X_t^{C1}}{MVE_{t-1}} + \beta_2 \cdot \frac{(X_t^{C1} - X_{t-1}^{C1})}{MVE_{t-1}} + ... + \beta_1 \cdot \frac{X_t^{Ci}}{MVE_{t-1}} + \beta_2 \cdot \frac{(X_t^{Ci} - X_{t-1}^{Ci})}{MVE_{t-1}} + \varepsilon_t .$$

The components, Ci, refer to the previously defined performance metrics, as mentioned yet above: CFO is the first component or base C_1, Accr is the second component C_2, ATInt is the third component C_3, and CapChg is the fourth component

$$q_{01} = \frac{\frac{n}{2}\ln\left(\frac{s_Z^2}{s_X^2 + \frac{1}{N}b'X'M_zXb}\right)}{\sqrt{\left(s_X^2 + s_Z^2 + \frac{1}{N}b'X'M_ZXb\right)^2 - \frac{s_X^2}{b'X'M_ZM_XM_ZXb}}}$$

where s_X^2 is the mean squared residual from the regression on y to X, s_Z^2 is the mean squared residual from the regression on y to Z, $b'X'M_ZXb$ is the sum of squared residuals from the regression of $X*b$ to Z, and $b'X'M_ZM_XM_ZXb$ is the sum of squared residuals from the regression of residuals from the regression of $X*b$ to Z to X.

[398] See Fidell and Tabachnick (2001, pp. 144-145)
[399] See White (1980).

C_4. For CRI, DepAdj[1] follows at final component C_5. For CEVA, AccAdj go after as fifth component C_5 and DepAdj[2] as final component C_6.

Again, H_2 and H_4 additionally involve two-tailed tests of the null hypotheses that unique components of CEVA do not provide incremental information in addition to that contained in operating cash flows. Rejection of H_2 or H_4 provides evidence of significant incremental information content. Significant difference in incremental information content is measured via standard F-tests.

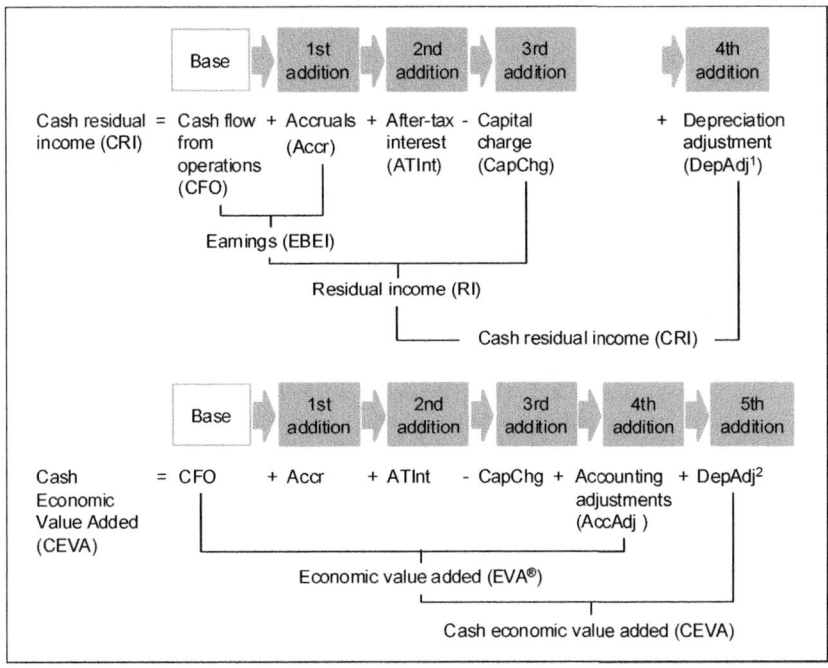

Figure 22   Decomposition of Newly Introduced CRI and CEVA Metrics

4.2.4   Sample Selection Process Gaining Panel Data

The sample consists of 201 U.S. firms,[400] representing both adopters and non-adopters of VBM. These firms exclusively fulfill the following criteria: They all represent U.S. firms with primary or secondary equity quotations at the New York Stock Exchange that belong to the general industrials, consumer goods (except tobacco), consumer services, or industrial transportation sectors and reported sufficient and consistent predefined data for at least five fiscal years within the research period 1995-2006. These firms, which all follow US-GAAP, provide 2,147 firm-year observations for fiscal years 1995 through 2006.

Within the selected sectors, firms were primarily chosen on the basis of availability of sufficient data from Thomson Financial market and accounting files. Initially, all primary and secondary equity quotations of dead, suspended and active equities at the New York Stock Exchange, filed by Thomson Financial as of January 18, 2007, were considered that referred to one of 11 sectors, as defined by the Industry Classification Benchmark of the New York Stock Exchange.[401] Then, out of 2,959 firms, 2,622 firms did not provide sufficient data for the purpose of this study.[402] Next, 325 firms provided less than five annual observations. Finally, 12 firms implied inconsistent, dual, or non-standard data and 25 firms did not fulfill geographic restrictions.[403]

Although the sample is limited to listed U.S. firms according to US-GAAP, derived implications may be equally valid for firms that are privately held, domiciled in another mature market, or following comparable accounting standards such as IFRS.

As shown in the subsequent figure, three distinct features characterize the underlying sample, as opposed to other samples adopted from league tables:

- First, a minimum of five observations per firm was viewed as a reasonable compromise between retaining enough firms, having sufficient observations per firm

---

[400] See appendix, Table 36.

[401] 28 excluded sectors include oil & gas producers, oil & gas equipment/services/distribution, chemicals, mining, industrial metals, forestry & papers, construction & materials, aerospace and defense, electronic & electrical equipment, industrial engineering, industrial support services, automobiles & parts, tobacco, health care equipment/services, pharmaceuticals & biotechnology, fixed line communications, mobile communications, electricity, gas/water/multi-utilities, banks, non-life insurance, life insurance, real estate, general financial services, equity services, non-equity services, software/computer services, technology/hardware/equipment.

[402] 2,520 firms with missing required Worldscope data items and, subsequently, 102 firms with missing Datastream market data; required Worldscope items consist of current-year net cash flow - operating activities (WC04860), net earnings before extraordinary items and preferred dividends (WC01551), interest expense on debt (WC01251), and operating income (WC01250); current-year and prior-year total assets (WC02999), total debt (WC03255), and total liabilities (WC03351); and current and four prior years extraordinary charge - pretax (WC01254), extraordinary credit - pretax (WC01253), extraordinary items & gain/loss sale of assets (WC01601), and non-operating interest income (WC01266); required Datastream items include return index (RI) and market value (MV).

[403] Four firms showed a minimum change of one month in the fiscal year end date, seven firms showed multiple listings in addition to the prevalent stock, one firm showed non-standard financial data, and 25 firms were not headquartered in the United States.

and minimizing the volatility in observations by fiscal year. Following, observations equally represent fiscal years. On the contrary, samples derived from league tables presumably reflect latest years overproportional.

- Second, the sample includes dead and inactive New York traded equities (regardless of the trading status), to reduce survivorship bias. Active, dead and suspended firms accounted for 86%, 12% and 2% of observations, respectively. [404] In contrast, samples based on league tables include active equities only.

- While the Stern Stewart Performance 1000 ranking excludes only real estate and financial services firms, firms in the sample are from non-financial and relatively mature and stable sectors, to minimize sector-induced distortions from exogenous factors, not controllable by management, and to avoid differing accounting schemes across sectors that make it difficult to generate strictly comparable data. Based on 39 sectors defined by the Industry Classification Benchmark, firms refer to 11 sectors from the industrials, consumer goods, and consumer services industries. Additionally, 25 non-U.S. firms referring primarily to immature markets and/or imposing bias from differing accounting standards are excluded. [405]

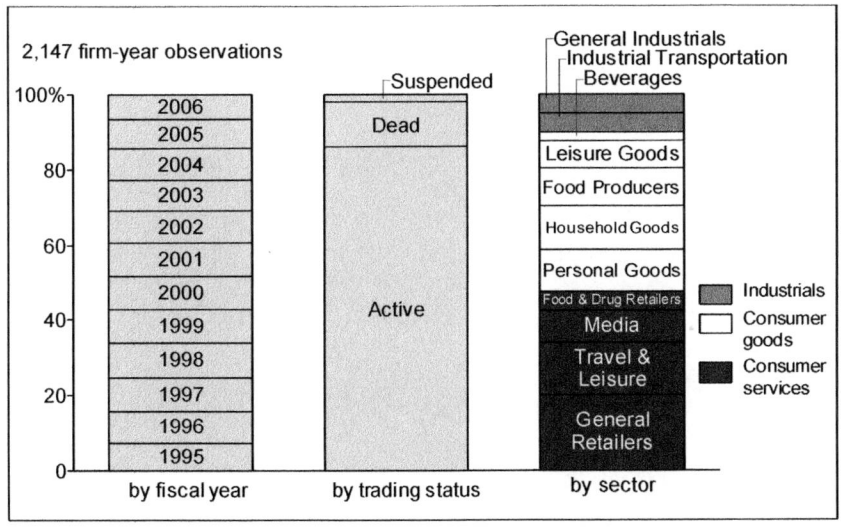

Figure 23   Sample Composition

---

[404] Trading status as of January 18, 2007.
[405] 24 American Depository Receipts and one Global Depository Receipt issued by firms from the following markets: Mexico (8), Chile (7), Australia (2), Italy (2), Netherlands (2), United Kingdom (2), China (1), and Luxembourg (1).

However, application of all 2,147 observations commonly results in biased regression results that are driven by very few outliers.[406] Therefore, extreme values above or below 8 standard deviations from the median are excluded. Depending on the underlying model specification and variables, the sample includes 11 to 24 annual observations from outliers.

After excluding extreme values, associations of accounting performance with market value were tested relatively with 2,136 observations and incrementally with 2,128 observations. Associations with returns, requiring the current and lagged firm performance level, were examined based on 1,925 and 1,922 observations for relative and incremental tests, respectively.

---

[406] E.g., O'Byrne and Young (2001, pp. 36-40) examine associations of scaled EVA, i.e. EVA divided by capital, with scaled MVA, using ranking data provided Stern Stewart & Co. for the 1,000 largest capitalization U.S. firms: They find that R^2 decreases from 27% to 7% after excluding observations from one firm, namely Wellpoint Health Network.

4.2.5   Definition of Independent and Dependent Variables

In this study, values of independent and dependent variables are set using Worldscope and Datastream database items; in some instances, corresponding Compustat items provide comparability to preceding studies.[407]

Independent Variables Describing the Firm's Overall Performance

In relative information content tests,[408] the independent variable is annual operating performance of a firm, quantified via six differing accounting measures: Initially, two traditional measures, namely an accrual-based earnings measure and a cash flow measure, provide a benchmark for value-based measures and consistency to traditional association studies, examining how key accounting measures explain stock prices.[409] Additionally, four residual income measures aim to quantify additions to shareholder wealth in the view of VBM and, thus, represent the focus of the study. Consequently, this study analyzes the following six performance measures:

- Cash Flow from Operations (CFO)

- Earnings before Extraordinary Items (EBEI)

- Residual Income (RI)

- Cash Residual Income (CRI)

- Economic Value Added (EVA)

- Cash Economic Value Added (CEVA)

Cash Flow from Operations (CFO)

CFO, obtained via Worldscope item WC04860, represents net cash receipts and disbursements resulting from the firm's operations, including extraordinary items.[410] By definition, CFO excludes all financing cash flows, such as common and preferred dividends. Applied data is consistent with data obtained via Compustat item D308, used by the Biddle et al. study.[411]

---

[407] All definitions of Worldscope items are obtained from the Thomson Financial 'Worldscope database datatype definitions guide', issue 5, December 2003. All definitions of Compustat data items are obtained from the 'Standard & Poor's Compustat (North America) user's guide', 2001. For Datastrem items, Thomson Financial does not provide an equivalent definition guide.

[408] See section 4.2.3.

[409] Value-relevance of earnings and cash flow has been examined by numerous authors (see Ali & Pope, 1995; Ball & Brown, 1968; Bernard & Stober, 1989; Biddle & Seow, 1991; R. M. Bowen et al., 1987; G. P. Wilson, 1986, 1987).

[410] Categorization of operating cash flows follows FAS 95.

[411] See Biddle et al. (1997, p. 312).

Earnings before Extraordinary Items (EBEI)

Retrieved via Worldscope item WC01551, EBEI equals net earnings after operating and non-operating income and expense, reserves, income taxes, minority interest and equity in earnings. Therefore, earnings data is consistent to data retrieved via Compustat item D18, which is frequently used.[412]

Contrarily to CFO, definition of EBEI excludes not only preferred and common dividends, but also any after-tax extraordinary items, such as effects from changes in accounting principles. However, EBEI includes interest on debt, minority interest and pretax extraordinary items, such as restructuring charges. Therefore, EBEI represents 'ordinary' earnings to all shareholders except shareholders of subsidiaries of the firm.

Residual Income (RI)

RI represents the most basic value-based measure, equaling net earnings after charges for all sources of net capital.[413] While EBEI only includes interest expenses for debtholders, RI implies also deductions for common and preferred stock.

For the following study, RI is defined as earnings before after-tax interest expense (i.e. earnings before extraordinary items, *EBEI*, and interest expense on debt before capitalized interest, *IntExp*, after taxes) less the weighted average cost of capital, *WACC*, times average net assets, *NA*, excluding non-interest bearing current liabilities, *NIBCL*, as follows: [414]

$$RI_t = EBEI_t + IntExp_t \cdot (1-t) - WACC_t \cdot \left(\frac{NA_{t-1} + NA_t}{2} - \frac{NIBCL_{t-1} + NIBCL_t}{2} \right).$$

EBEI, IntExp, and NA are obtained from Worldscope items WC01551, WC01251, and WC02999, respectively. WACC and NIBCLs estimates are taken from EVA calculations.

Although valuation theory requires using beginning capital, the definition above includes average capital. The underlying rationale is to ensure consistency between periodic earnings and average capital used to generate these returns. Further, any tax effects are calculated based on effective tax rates.[415] However, in years with negative

---

[412] See Biddle et al. (1997, p. 313), Biddle and Seow (1991, p. 199), Bowen et al. (1987, p. 729), and Copeland et al. (2004, p. 15).

[413] See section 2.2.2.

[414] The EBO model defines RI based on net earnings, beginning of period capital and the cost of equity (Edwards & Bell, 1961; Ohlson, 1990, 1995). The definition applied in this study aims to use an earnings measure consistent with the traditional earnings measure, to provide consistency between earnings and the capital charge, and to apply the same capital charge as in EVA computations. Underlying, the study focuses on annual performance measurement rather than assessment of market prices via the EBO model, in terms of a multi-period security valuation model.

[415] Going forward, effective tax rates, approximated by dividing income tax expense (WC01451) by pretax income (WC01401), are used. Consequently, effective tax estimates include federal and state income tax rates, firm-specific tax effects such as loss carryforwards, international operations, and restructuring charges, changes in tax rates over time as well as differing tax rates across countries and states.

pretax income, the tax rate is assumed to be zero. The following measures are equally defined based on average capital and effective tax rates.

Cash Residual Income (CRI)

CRI aims to overcome age bias and primarily differs from RI by replacing book depreciation with economic depreciation, *ED*, and by using gross rather than net earnings and capital. Specifically, CRI is defined as earnings before extraordinary items, *EBEI*, interest expense on debt before capitalized interest, *IntExp*, and depreciation and amortization expense, *DA*, less ED and the weighted average cost of capital, *WACC*, times the average net assets, *NA*, and accumulated depreciation and amortization, *ADA*:

$$CRI_t = \left(EBEI_t + IntExp_t \cdot (1-t) + DA_t\right) - ED_t - WACC_t \cdot \left(\frac{NA_{t-1} + NA_t}{2} + \frac{ADA_{t-1} + ADA_t}{2}\right).$$

Economic depreciation is determined by gross depreciable assets, *GDA*, the assets' useful life, *T*, and the cost of capital, *WACC*, as follows:

$$ED_t = GDA_{t-1} \cdot \frac{WACC}{(1+WACC)^T - 1}.$$

GDA shall equal to the sum of net plant, intangibles, and ADA less the sum of land, construction in progress, gross goodwill, and other non-amortizable intangibles. T may be approximated dividing beginning gross depreciable assets, *GDA*, by the depreciation and amortization expense, *DA*. EBEI, IntExp, DA, NA, net plant, and (other) intangibles are retrieved from Worldscope items WC01551, WC01251, WC01151, WC02501, WC02649 and WC02999. ADA, land, construction in progress, gross goodwill, and other non-amortizable intangibles are obtained from individual annual filings, in particular from notes titled 'Property and Equipment' and 'Goodwill and Other Intangible Assets';[416] WACC estimates are used from EVA calculations as cost of capital and reinvestment rate, assuming that capital weights do not alter due to the conversion of net capital into gross capital and current costs of capital are expected to be constant.

The formula of economic depreciation apparently follows the CVA concept, but is actually derived from a different theoretical approach aimed at ensuring mathematical reconciliation of CRI with the RI and DCF valuation theories. In particular, economic depreciation applied in this study differs from economic depreciation according to the CVA method, as follows:[417]

• Depreciable assets are not adjusted for inflation. This also avoids subjective inflation estimates.

---

[416] If financial statements provide no separate carrying amounts of land and buildings, the combined amount is treated as 'land' in the estimation of the economic depreciation, assuming both do not deteriorate considerably. Furthermore, if financial statements provide net rather than gross goodwill, accumulated goodwill amortization is additionally deducted, if disclosed.

[417] For a detailed calculation of the CVA and CFROI metric, see Madden (1999) and Martin and Petty (2000, p. 236).

- Depreciable assets also include straight-line amortizing intangible assets, such as capitalized software development expenses, patents, copyrights, leasehold improvements and trademarks, which introduce age bias equivalent to tangibles.[418]
- Construction in progress (CIP) is excluded from depreciable assets until placed in service.
- Non-capitalized operating leases are excluded from depreciating assets, since interest on leases yet accounts for deterioration in value.
- Additionally, goodwill is excluded from depreciating assets, assuming goodwill does not deteriorate.
- Finally, the WACC replaces the so called reinvestment rate. This also avoids complexity of CFROI calculations.

Economic Value Added (EVA)

EVA represents a more complex residual income measure that includes a wide series of accounting adjustments to net earnings and net capital, defined as net operating profits after taxes, *NOPAT*, less the weighted average cost of capital, *WACC*, times the average capital:[419]

$$EVA_t = NOPAT_t - WACC_t \cdot \left(\frac{Capital_{t-1} + Capital_t}{2} \right).$$

Differing from EVA data published by Stern Stewart & Co., [420] this study independently estimates EVA data, as follows:[421]

NOPAT = *EBEI* + non-operating income (expenses) + after-tax unusual losses (gains) + financing costs + after-tax lease interest + Δ deferred taxes + + Δ reserves + after-tax goodwill amortization + Δ non-capitalized intangibles + pension adjustment + ESOP contribution + fair value reserve.

Capital = *NA* + accumulated after-tax unusual losses (gains) − non-operating assets − *NIBCLs* + present value of operating leases + reserves + cumulative goodwill amortization + non-capitalized intangibles + net pension liability.

$$WACC_t = \frac{D_{t,t-1}}{Capital_{t,t-1}} \cdot \left[\frac{IntExp_t}{Debt_{t,t-1}} \cdot (1 - t_t) \right] + \frac{E_{t,t-1}}{Capital_{t,t-1}} \cdot (8.6\% + \beta_t \cdot 6\%).$$

---

[418] SFAS 142 requires straight-line amortizing intangible assets with a finite useful life and an unknown pattern, in which the respective economic benefits are consumed.

[419] See section 2.2.2.

[420] E.g. EVA data provided by Stern Stewart implies differing accumulation periods for goodwill amortization or tax rate assumptions, see Stewart (1991, p. 744): Goodwill amortization is accumulated since 1973. Moreover, general marginal tax rate assumptions are applied for all firms by fiscal year (1991, p. 743).

[421] For an overview of the introduced calculation methodology, see appendix, Table 34 and Table 35.

The above definitions of *NOPAT* and *Capital* include the following adjustments to net earnings, *EBEI*, and net assets, *NA* (the latter obtained via Worldscope items WC01551 and WC02501, respectively):

- While *EBEI* already excludes after-tax unusual items, capital is adjusted by adding (subtracting) cumulative unusual losses (gains) to (from) net assets. Data on extraordinary items is derived from Worldscope items WC01253, WC01254 and WC01601 and accumulated starting in 1991.[422] Pretax extraordinary items are multiplied with 1 minus the effective income tax rates.

- To adjust for non-operating assets, construction in progress (CIP), obtained from the financial statement note on 'Property, Plant and Equipment', and other investments, obtained from Worldscope item WC02250, are excluded from capital. Adjustments for non-operating holdings of treasury stock are not necessary, since examined firms follow US-GAAP.[423] Moreover, cash and cash equivalents, excess cash, and equity investments are assumed to be held for business activities, since adjustments would mostly require subjective judgments;[424] in addition, there is no consistent and objective criterion to validate that non-consolidated investments accounted for under the equity method are operational.[425]

- The average beginning and end of year average of non-interest bearing current liabilities, *NIBCLs*, obtained via Worldscope items WC03051 and WC03101, is subtracted from net assets.

- Non-operating income (expenses) is subtracted from (added back to) *EBEI*. Interest income data is obtained from Worldscope item WC01266, pretax non-operating income and expenses from items WC01253, WC01254, WC01266 and WC18191, and after-tax non-operating income and expenses from items WC01451, WC01501 and WC01551.

---

[422] EVA data provided by Stern Stewart & Company includes accumulated capital adjustment for unusual losses and gains since 1973 or since publicly data became available (G. B. Stewart, 1991, p. 744), an accumulation period imposing difficulties for investor communication.

[423] US-GAAP require subtraction of own stock from stockholders' equity as treasury stock, see ARB 43 ch. 1A. Otherwise, treasury stock had required to be excluded, since investments in own stock are not considered strategic, do not carry an asset value at the time of liquidation and may be used to manage the share price, see Hostettler (1997, pp. 118-120).

[424] See Stewart (1991, p. 744) and Copeland et al. (2000, p. 161), both requiring to exclude non-operating cash and equity investments. US-GAAP accounts for marketable securities following SFAS 115, 80 and 133. Cash invested in non-operating marketable securities is included in 'Cash and Cash Equivalents' but not always separately disclosed. Also, excess cash not used for daily operations, is not disclosed. Therefore, Copeland et al. (2000, p. 161) suggests adjusting for excess cash based on a reliable estimate of target cash balances by industry.

[425] See Wolin & Klopukh (2000, p. 148); Hostettler (1997, p. 116) classifies the contractual cooperation of Swissair and Sabena and strategic R&D, production and manufacturing investment of Ciba in Chiron as operating equity investments, proposing to use, e.g. contractual obligations or relevance to the strategic positioning of the firm, as evidence for operating investments.

- While *EBEI* is yet defined before common and preferred dividends, other financing costs such as minority interest and interest on debt are to be added back to earnings. Minority interest is obtained from Worldscope item WC01501 and interest expense on debt less capitalized interest is obtained via Worldscope items WC01251 and WC01255. Effective income tax rates are used to estimate the after-tax effect of net interest expense.

- Accounting for non-capitalized operating leases, the present value of the five future minimum rentals at 10% is added to capital, representing a debt equivalent, and the after-tax interest component at 10% on the average capitalized value is added to earnings, representing a figurative cost of debt.[426] Rentals are assumed to be paid in the middle of the year. After-tax interest expense is estimated based on effective income tax rates. Data on rentals is obtained from annual filings. If only the aggregate amount of minimum rental payments over years one to five is provided, lease payments are assumed to be constant over time.

- All reserves that are disclosed in the financial statements, such as the Last-in-First-out (LIFO) reserve[427] or other precautionary reserves[428], are reversed. Judgments on the substantiality and volatility of reserves are not applied.[429]

- To reverse write-down of goodwill, average beginning and end of year cumulative goodwill amortization before tax is added to capital. Additionally, after-tax goodwill amortization, estimated based on effective tax rates, is added to earnings. Data on goodwill amortization is obtained from notes to the financial statements. Reversal of goodwill impairments is unnecessary, since yet included in the adjustment for un-usual losses. Unrecorded goodwill is not capitalized, to avoid subjective categorizations of mergers and information not readily available from annual filings.

- Research and development (R&D) expenses, obtained via Worldscope item WC01201, and marketing expenses, obtained from financial statements, are capitalized straight-line over 5 years and 2 years, respectively, as intangibles.[430]

---

[426] To ensure consistency with EVA data provided by Stern Stewart & Company, see Stewart (1991, pp. 98-99)., information provided on future lease payments beyond year five or on the interest rate on lease obligations or debt, see   Wolin and Klopukh (2000, p. 145) and Peterson and Peterson (1996, p. 15), is disregarded.

[427] Regarding the LIFO reserve, the EVA concept requires to value assets at market value at the end of period with any gains classified as current period income to reflect the nominal cost of capital, see Stewart (1991, p. 113).

[428] Precautionary reserves are added to capital and the yearly increase or decrease in reserves is added to or subtracted from earnings, see Stewart (1991, p. 117).

[429] Stewart (1991, p. 117) suggest to reverse reserves only if growing relative with the level of business activities. Copeland et al. (2000, pp. 164-165) advocate to reverse reserves only in the case of a substantial deviation between market and book values, as for long-term assets such as real estate and airplanes.

[430] Intangibles are to be capitalized over the anticipated payoff period. The payoff period for R&D projects is commonly assumed to be 5 years, see Stewart (1991, p. 190). The payoff for marketing

Correspondingly, increases or decreases in net intangibles are added to or subtracted from earnings, respectively. Other expenses are not capitalized due to insufficient disclosure requirements.[431]

- Non-current deferred taxes, obtained via Worldscope item WC03263, are reclassified as equity rather than debt and non-cash expense increasing non-current deferred taxes is added back to earnings.[432]

- Pension and postretirement obligations less the fair value of plan assets are capitalized and the portion of pension and postretirement expense that does not refer to interest cost is added back to earnings.[433] Data is attained from financial statements.

- Capital costs on 'Employee Stock Ownership Plan' (ESOP) loans, i.e. contributions to the ESOP to service principal and interest on the ESOP loan, are deducted from earnings as compensation expense.[434] Data on cash contributions is gained from annual filings.

- If stock-based compensation expense is not recognized under the fair value method for fiscal years prior to 2006, stock option expense is adjusted to the fair market value of the option on the date of grant.[435] Data on fair value reserve of stock-based compensation expense is obtained from financial statements.

---

costs over the lives of successful new products is assumed to be 2 years, see Wolin and Klopukh (2000, p. 146).

[431] The EVA concept suggests capitalizing and amortizing any expenses with long-term payoff periods, see Stewart (1991, pp. 115-116, 185, 190, 742, 2003, p. 75).

[432] See Stewart (1991, p. 190). I avoid subjective verification that deferred taxes occur from ongoing timing differences related to operating activities and apply reverse adjustments to non-current deferred tax assets. Other authors support that the adjustment for deferred taxes also applies to non-current deferred tax asserts (see Hostettler, 1997, p. 129; Peterson, 2000; Wolin & Klopukh, 2000).

[433] See Stewart (2003, pp. 69-72).

[434] See Stewart (1991, p. 534).

[435] See Stewart (2003, p. 81).

Finally, WACC calculation follows mostly the approach of Stern Stewart, to ensure consistency with studies applying data provided by Stern Stewart, as follows:

- The pretax cost of debt is estimated by dividing gross interest expense before capitalized interest by the average beginning and end of year balance of interest-bearing debt. The after-tax cost, including the tax shield of debt, is estimated via firm-specific effective income tax rates. The market cost of debt is not applied, since respective market data is not consistently publicly available or require undisclosed rating methodologies.[436] Interest expense, short term and long term interest-bearing debt, are retrieved from Worldscope items WC01251, WC03051 and WC03251.

- The cost of equity is derived from the 'Capital Asset Pricing Model' (CAPM), based on a risk-free rate of return of 8.6%, a market risk premium of 6% and a beta factor annually derived by regressing monthly common stock returns of the firm against monthly returns of the S&P 500 for five precedent years.[437] Betas are aggregated by ICB sector, to minimize distortions from the CAPM.[438] Monthly return data is obtained from Datastream item RI, representing closing prices at the last trading day of the month.

- Capital weights are based on the average beginning and end of fiscal year book value of debt, preferred stock, and minority interest and average daily market value

---

[436] See Stewart (1991, pp. 434, 392). Stern Stewart & Company suggests deriving the market cost of debt from the yield to maturity on publicly traded debt of the firm or from the average yield to maturity of bonds of firms with equivalent bond rating. Bond ratings are estimated via the Stern Stewart & Co. Bond Rating Scoring System, which is primarily based on five factors: Size, risk-adjusted return, ratio of long-term debt to total capital, ratio of adjusted total liabilities to net worth, ratio of investments and advances to unconsolidated subsidiaries to total capital; although these factors account for 70 percent of the bond rating, full transparency on the rating methodology is not provided.

[437] Stewart (1991, pp. 436-442, 743); Stern Stewart & Company obtain the market risk premium from a CRSP study that examined the difference in annual rates of return from the Standard & Poor's 500 stock index and 20-year U.S. Treasury Bonds for the years 1925 to 1989; Copeland et al. (2000, pp. 260-261) find a similar level of market risk premium of 5% derived from the difference between the geometric mean return on the market and long-term government bonds over 1926 to 1993. Copeland et al. (2000, pp. 260-261) recommends using a risk-free rate of return from 10-year U.S. Treasury bond returns. The risk-free rate of 8.6% was derived by Simon and Hunter (2005) who calculated weekly 10-year U.S. Treasury bond returns by multiplying the change in rates by the modified duration and adding accrued interest earned for the period of 1992 to 2002. As another reference, Stewart (1991, p. 442) refers to long-term U.S. government bond yields to maturity of 8.8%.

[438] Latest empirical research does not support the CAPM, see Fama and French (1992); although application of the arbitrage pricing theory or other approaches (e.g. fundamental factor models such as BARRA and Fama-French, macro-factor models such as Burmeister, Ibbotson, Roll, and Ross and Salomon RAM, the EVA factor model or analysts' estimates) may yield estimates closer to the expected return on equity, these models are considered too complex and insufficiently known to investors to be recognized in market data. However, empirical results suggest that estimation of the cost of capital via the CAPM or the Fama-French model does not have a notable influence, the CAPM model represents a well-known and straight-forward method to determine the cost of equity, see Tsuji (2006).

of common equity.[439] Book values of debt, preferred stock and minority interest are obtained via Worldscope items WC03051, WC03251, WC03451 and WC03426. The daily market value of equity is derived from Datastream items P and NOSH.

Cash Economic Value Added (CEVA)

CEVA represents an EVA measure based on gross adjusted earnings and gross adjusted capital, including an additional depreciation adjustment to reduce age bias. CEVA is computed as follows:

$$CEVA_t = (NOPAT_t + DA_t - GA_t) - ED_t -$$

$$-WACC_t \cdot \left(\frac{Capital_{t-1} + Capital_t}{2} + \frac{ADA_{t-1} + ADA_t}{2} - \frac{AGA_{t-1} + AGA_t}{2} \right).$$

$NOPAT$, $Capital$, GA = periodic goodwill amortization, AGA = cumulative goodwill amortization, and $WACC$ are all obtained from EVA calculations. ED = economic depreciation (see definition above), DA = periodic depreciation and amortization, and ADA = cumulative depreciation and amortization are obtained from CRI calculations.

Incremental Information Components Describing the Firm's Performance

In incremental information content tests,[440] the independent variables are analyzed in a sequence in terms of regression relevance of incremental information components. In this study incremental analysis refers to operating cash flows as base and analyzes the following information components:

- Accruals (Accr)

- After-Tax Interest Expense (ATInt)

- Capital Charge (CapChg)

- Accounting Adjustments (AccAdj)

- Depreciation Adjustment (DepAdj)

Accruals (Accr)

Accruals, *Accr*, are unique information contained in EBEI beyond that contained in CFO. Consequently, accruals are defined as EBEI, obtained via Worldscope item WC01551, less CFO, obtained via Worldscope item WC04860, notationally:

$Accr_t = EBEI_t - CFO_t.$

Accruals include depreciation, amortization, change in non-cash current assets, change in non-current portion of deferred taxes, and change in current liabilities other than notes payable and the current portion of long-term debt, and non-operating income less after-tax extraordinary income. Accruals are contained in all following residual income measures, such as RI, CRI, EVA, and CEVA.

---

[439] Capital weights based on the target capital structure at market values over the trailing three years (see G. B. Stewart, 1991, p. 743) cannot be derived from publicly available information.

[440] See section 4.2.3.

After-Tax Interest Expense (ATInt) / Capital Charge (CapChg)

After-tax interest expense, *ATInt*, and capital charges, *CapChg*, represent two incremental components uniquely contained in RI versus other information yet contained in EBEI, notationally:

$$ATInt_t + CapChg_t = RI_t - EBEI_t = RI_t - (CFO_t + Accr_t).$$

After-tax interest expense equals 1 minus the firm's effective tax rate times gross interest expense, obtained via Worldscope item WC01251. The capital charge reflects the weighted average of the cost of debt, preferred stock and minority interest, derived from book values, and the sector-specific CAPM-based cost of equity multiplied by average interest-bearing capital. Altogether, interest and capital charges are similarly included in any other residual income measure, such as CRI, EVA, and CEVA.

RI adds back interest on debt while charging for all sources of capital. Consequently, the ATInt component is predicted to be positive and the CapChg component is expected to be negative. To allow a more differentiated analysis, the two information components are examined separately.

Accounting Adjustments (AccAdj)

Accounting adjustments, *AccAdj*, represent information contained in EVA beyond that contained in RI, notationally:

$$AccAdj_t = EVA_t - RI_t = EVA_t - (CFO_t + Accr_t + ATInt_t + CapChg_t).$$

As described in the previous section, accounting adjustments imply several corrections to net earnings and net capital that refer to minority interest, non-capitalized operating leases, interest income, other non-operating income, employee stock ownership plans (ESOP), pension, postretirement and stock option plans, deferred taxes, Last-in-First-out (LIFO) reserve, bad debt reserve, deferred tax asset valuation allowance, other allowances, research and development expenses, advertising expenses, goodwill amortization, unusual losses or gains, construction in progress, other investments, accounts payables, and current accrued liabilities.

Depreciation Adjustment (DepAdj)

In this study, CRI and CEVA represent two VBM metrics adjusted for accounting depreciation. However, the respective adjustment slightly differs, since the underlying RI and EVA measures deal differently with goodwill amortization.

The depreciation adjustment of CRI, $DepAdj^1$, reflects information distinctively contained in CRI versus other information contained in RI, notationally:

$$DepAdj_t^1 = CRI_t - RI_t = CRI_t - (CFO_t + Accr_t + ATInt_t + CapChg_t).$$

The depreciation adjustment of CEVA, $DepAdj^2$, corresponds to unique information added by CEVA versus information yet conveyed by EVA, notationally:

$$DepAdj_t^2 = CEVA_t - EVA_t = CEVA_t - (CFO_t + Accr_t + ATInt_t + CapChg_t + AccAdj_t)$$

$DepAdj^1$ equals book depreciation and amortization from Worldscope item WC01151 less economic depreciation and a capital charge on accumulated depreciation and amortization. Although $DepAdj^2$ is largely equivalent to $DepAdj^1$, it excludes goodwill amortization yet considered within AccAdj.

Dependent Variables Describing the Firm's Market Performance

The study assumes a causal relationship between accounting performance and shareholder wealth. According to valuation theory and conventional accounting research, two market measures of firm performance serve as dependent variables:

- Market Value of the Firm (MV)

- Abnormal Stock Returns (AR)

Market Value of the Firm (MV)

'Market Value' (MV) of the firm is defined as the sum of the firm's average daily market value of equity, MVE, and average book value of net capital, NA, less its average book value of common equity, CE, over the fiscal year of the firm, notationally:[441]

$$MV_t = MVE_t + \left(\frac{NA_{t-1} + NA_t}{2} \right) - \left(\frac{CE_{t-1} + CE_t}{2} \right).$$

The arithmetic average of the daily market value of common stock over the fiscal year is obtained from Datastream items P and NOSH, net assets from Worldscope item WC03999, and common equity, including common stock value, retained earnings, capital surplus and capital stock premium, via Worldscope item WC03501. Implicitly, the definition assumes that market valuations of minority interest, preferred stock and debt capital do not considerably differ from book values.

Instead, some studies use an excess market value of the firm, termed 'Market Value Added' (MVA), as dependent variable.[442] MVA is defined as *"difference between a firm's market value and its capital employed"*.[443] However, replacement of firm value with MVA as dependent variable does not affect regression results perceptibly.

Abnormal Stock Returns (AR)

'Abnormal Returns' (AR) represent a proxy of unsystematic returns. Computed as arithmetic market-adjusted returns, AR equal monthly returns to common shareholders of firm i, r_i, less monthly returns from the value-weighted MSCI U.S. Investable Market 2,500 index, r_m, which approximate market-wide returns.

---

[441] Stewart (1991, p. 741) uses excess market value, also referred to as Market Value Added (MVA), namely the difference between market value and employed capital, as dependent variable. However, I apply total market value (see Biddle et al., 1997; O'Byrne, 1996).

[442] See Fernández (2001, 2002), Garg and Singh (2004), Heidorn et al. (2000), Kramer and Pushner (1997), Stewart (1991), and Ramana (2007).

[443] See Stewart (1991, pp. 153, 741).

Notationally, AR are defined as follows (where τ = index aligned with firm's fiscal year, e.g. τ = 1 for the first month of the 12-month period):[444]

$$AR_{i,t} = \Pi_{\tau=1}^{12}(1 + r_{i,\tau}) - \Pi_{\tau=1}^{12}(1 + r_{m,\tau}) - 1.$$

This definition implies that index returns are matched to the fiscal year of the respective firm. To avoid look-ahead bias, the 12-month period ends three months after the end of the fiscal year-end month, consistent to SEC (U.S. Securities and Exchange Commission) filing requirements for annual reports. [445] Return data is obtained from Datastream item RI, which includes capital gains plus dividend returns adjusted for all capital distributions to shareholders, e.g., stock splits and dividends, spin-offs, and repurchases or issuances of equity.[446] Selection of the value-weighted MSCI index as reference portfolio, representative for large U.S. stock, avoids misspecifications due to new listings, rebalancing, and skewness bias as prevalent with equally-weighted indices; [447] further, it circumvents matching the firm via market capitalization or book-to-market-ratios;[448] finally, it evades construction of a reference portfolio.[449]

---

[444] Several studies apply abnormal returns (Biddle et al., 1997; Biddle et al., 1995; R. M. Bowen et al., 1987). The MSCI US Investable Market 2500 index represents about 98% of the capitalization of the US equity market (see http://www.mscibarra.com). I avoid using the S&P 500 index that does not include dividend income (see Standard & Poor's "S&P 500" factsheet, December 31, 2006).

[445] The assumption of reported information to be impounded in market prices after three months is commonly used (see Biddle et al., 1997; Easton & Harris, 1991; Fama & French, 1992; Peterson & Peterson, 1996; Rayburn, 1986). Further, Bowen et al. (1987, p. 732) found for 98 firms 10-K filings were available no later than three months after the fiscal year end. Although they note that sometimes mailings have been deferred, the mailing date is not relevant since introduction of the electronic filing system. Historically, form 10-K had to be filed with the SEC (U.S. Securities and Exchange Commission) within 90 days after the end of the company's fiscal year (Securities and Exchange Commission. *Annual Report Pursuant to Section 13 or 15(d) of the Securities Exchange Act of 1934. General Instructions*). However, for fiscal years ending on or after December 15, 2003, a 60 day and 75 day filing date applies to large accelerated filers with a public float of at least USD 700 bn and accelerated filers with a public float of more than USD 75 bn and less than USD 700 bn, respectively, for fiscal years (Securities and Exchange Commission. RIN 3235-AI33: *Acceleration of Periodic Report Filing Dates and Disclosure Concerning Website Access to Reports*; Securities and Exchange Commission. RIN 3235-AJ29: *Revisions to Accelerated Filer Definition and Accelerated Deadlines for Filing Periodic Reports*).

[446] Datastream return index "RI" is defined as value growth of a share holding, if dividends are re-invested to purchase additional units of the underlying share at the closing price applicable on the ex-dividend date. The return index data refers to closing prices at the last day of the month; I calculate adjusted returns on the prevalent month of the fiscal year end for fiscal years 1997 to 1999.

[447] See Barber and Lyon (1996).

[448] See Barber and Lyon (1996) and Jacquillat and Solnick (1978).

[449] See Ritter (1991, p. 7) and Ritter and Loughran (1995, p. 36).

4.3 Summary

Numerous empirical studies examine the value-relevance of VBM metrics, i.e., the association of value-based measures with the market value of the firm or stock returns. Studies that are based on a sufficient sample of U.S. firms and a consistent study design show for the most part that the traditional earnings measure outperforms the value-based EVA measure, while cash flow seems to have even lower value-relevance. More recently, only some studies showed contrary evidence, finding that EVA is most value-relevant. Studies examining other measures than EVA provide some indications that REVA, TBR, and CVA are superior to EVA in explaining stock prices. However, no study allows comparing the value-relevance of a practicable and depreciation-adjusted VBM metric to the value-relevance of other common value-based and traditional metrics.

To overcome prior deficiencies in the definition of metrics and the conception of studies, the design of this study includes a sufficiently large and unbiased sample of U.S. firms, a broad and consistent regression approach, and independently calculated data. The sample consists of annual accounting results for fiscal years 1995 through 2006, of 201 listed firms acting in a broad range of industrials and consumer goods and services sectors, chosen within the selected sectors primarily upon data availability, providing 2,147 firm-year observations. The sample stands out by equally reflecting fiscal years, reducing survivorship bias, and reducing distortions from emerging, volatile and financials industries. Four research hypotheses relate to the relative and incremental information content of accounting measures of firm performance in their association with the firm's market value and stock market returns. Relative information content tests compare the value-relevance of operating cash flow, earnings, and independently computed RI, CRI, EVA, and CEVA. Incremental information content tests assess the additional information conveyed by earnings' accruals, the capital charge adjustment of RI, the accounting adjustments of EVA, and the depreciation adjustments of CRI and CEVA. Finally, this study integrates associations of performance measures with two different dependent variables, namely firm value and stock returns of the firm in excess of market returns, and provides an extensive sensitivity analysis.

5 Results

This chapter begins with a case study that illustrates defined independent and dependent variables, particularly newly defined CRI and CEVA metrics, aiding the comprehension of the underlying computation principles and rationales, and segueing into the following empirical study.

5.1 Case Study

In the following, variables that were defined for the empirical study are illustratively calculated for the U.S. food producer Campbell Soup Corporation, which represents an internationally well-known company that operates in a comprehensive and mature industry.[450] Campbell Soup has fairly middle-aged assets, as identified by the ratio of cumulative depreciation and amortization divided by total gross depreciable assets that is equal to 54%.[451]

To better detect age bias of accounting metrics, key data of Campbell Soup is compared with key data of two additional firms with substantially differently aged assets: the airline holding Alaska Air Group (with Alaska Airlines and Horizon Air as subsidiaries) and the uniform and work wear supplier Unifirst, which carry relatively new and old assets, respectively. On average, disclosed accumulated depreciation and amortization of Alaska Air Group accounts to 28% of its gross depreciable assets, while cumulative depreciation and amortization of Unifirst makes up 74% of its gross depreciating assets. All three firms are included in the sample of the following empirical study.

---

[450] Several authors provide illustrative calculations based on financial statement data from prevalent U.S. food producers such as Campbell Soup Corporation (G. B. Stewart, 1991) and Hershey Foods Corporation (Copeland et al., 2000; Madden, 1999; Peterson & Peterson, 1996). Contrary to Campbell Soup Corporation, Hershey Foods Corporation is not part of the sample.

[451] Data on accumulated depreciation and accumulated depreciation was obtained from individual annual filings of the firms. For the applied methodology to calculate gross depreciable assets, see section 4.2.5.

5.1.1   Accounting-Based Measures of Shareholder Value Creation

Six accounting measures were defined that all describe corporate performance: Cash flow from operations, earnings before extraordinary items, residual income, cash residual income, economic value added, and cash economic value added.[452]

Cash Flow from Operations (CFO)

The cash flow measure used in this study is 'Cash Flow from Operations' (CFO). Figure 24 shows CFO of the underlying period of the empirical study for the three selected companies: Average CFO is highest for Campbell Soup ($M = 1,088$ m USD), lower for Alaska Air ($M = 273$ m USD) and lowest for Unifirst ($M = 61$ m USD). While results for Campbell Soup and Unifirst are fairly constant, Alaska Air shows a substantial increase in operating cash flows that increase from 126 m USD in 1995 to 450 m USD in 2006. Therefore, the sizeable growth yet indicates that Alaska Air has relatively 'young' assets.

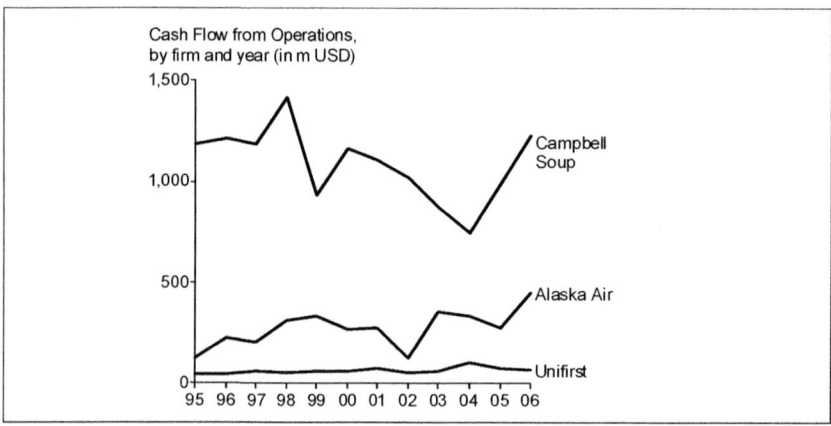

Figure 24   Case Study - Cash Flow from Operations

Earnings before Extraordinary Items (EBEI)

The applied earnings measure is 'Earnings before Extraordinary Items' (EBEI). As shown in Figure 25, the case example suggests that EBEI is generally considerably lower and less volatile than CFO. Average EBEI levels of Campbell Soup, Alaska Air, and Unifirst account to about 63%, 9%, and 47% of average CFO levels, respectively, providing evidence for the predominance of non-cash expenses. Compared with the standard deviation of CFO, standard deviation of EBEI of Campbell Soup, Alaska Air,

---

[452] See section 4.2.5.

and Unifirst is reduced by 62%, 29%, and 58%, respectively, providing evidence for the assumption that EBEI is substantially smoothed, for example by non-cash accounting accruals and reserves.

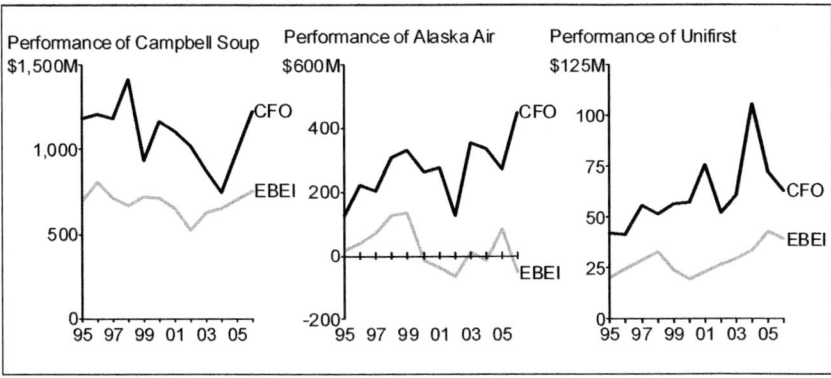

Figure 25   Case Study - Comparison of Cash Flow and Earnings

Residual Income (RI)

Calculation of 'Residual Income' (RI) is primarily based on three elements: Earnings before extraordinary items and after-tax interest expense, $EBEI_{BI}$, average interest-bearing net capital, *Net capital*, and the average weighted cost of capital, $WACC$; it is defined by the following equation:[453]

$$RI_t = EBEI_t + IntExp_t \cdot (1-t) - WACC_t \cdot \left(\frac{NA_{t-1} + NA_t}{2} - \frac{NIBCL_{t-1} + NIBCL_t}{2} \right).$$

Table 14 illustrates the RI calculation for Campbell Soup Corporation:

As shown in the first section of the table, adding back interest to earnings results in an average increase in earnings of 18%.[454] In the following section, non-interest bearing current liabilities (NIBCLs) are deducted from total net capital, reducing capital by on average 23%. Next, capital charges are deducted from earnings before interest, based on WACC estimates taken from EVA calculations (and, therefore, discussed later). As shown in the last row of the table, RI modestly increases from 260 millions USD in 1995 to 336 millions USD in 2006 (2.4% p.a.), resulting primarily from the reduction of the WACC from 12% to 10% (-1.8% p.a.).

---

[453] See section 4.2.5.
[454] Estimated effective tax rates, to exclude the tax shield of debt, range from 25% to 36%. Effective tax rates are used for various after-tax estimates within all independently computed residual income variables, to avoid errors due to general tax rate assumptions.

Year	EBEI	Interest expense	Effective tax rate	EBEI before interest	Average net assets	Average NIBCLs	Net capital	WACC	Residual income (RI)
1995	698	123	32%	781	5,654	-1,265	4,389	12%	260
1996	802	137	33%	894	6,474	-1,332	5,142	12%	284
1997	713	178	35%	828	6,546	-1,420	5,126	12%	211
1998	671	194	36%	796	6,046	-1,439	4,608	11%	274
1999	724	190	34%	849	5,578	-1,281	4,297	11%	378
2000	714	204	34%	849	5,359	-1,159	4,200	10%	445
2001	649	222	34%	795	5,562	-1,237	4,325	9%	406
2002	525	191	34%	651	5,824	-1,398	4,426	9%	274
2003	626	188	32%	753	5,963	-1,493	4,470	8%	375
2004	647	177	32%	768	6,440	-1,517	4,924	9%	343
2005	707	188	31%	836	6,723	-1,540	5,183	9%	353
2006	755	170	25%	883	7,320	-1,708	5,612	10%	336

Table 14   Campbell Soup Corp.: Residual Income (in m USD)

As shown in Figure 26, RI levels are consistently lower than EBEI levels, since RI charges for all sources of capital while earnings charges only for debt capital.

Moreover, Figure 26 clearly shows the conceptual advantage of RI over earnings, as follows. Positive earnings of all firms do not allow drawing any conclusions on value addition. Instead, positive or negative levels of RI can be interpreted as additions to shareholder value or destruction of shareholder wealth, with respect to the book value of net assets. Consequently, Campbell Soup consistently created shareholder value (as shown by positive RI levels) with respect to its net book value, while Alaska Air and Unifirst persistently destroyed shareholder value (as indicated by negative RI levels).

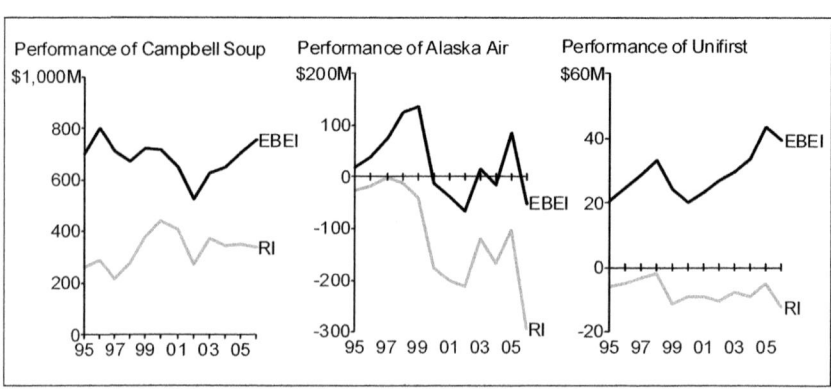

Figure 26   Case Study - Comparison of Earnings and Residual Income

Economic Value Added (EVA)

Calculation of 'Economic Value Added' (EVA) follows the equation:

$$EVA_t = NOPAT_t - WACC_t \cdot \left(\frac{Capital_{t-1} + Capital_t}{2} \right).^{455}$$

Table 15 outlines EVA computation for Campbell Soup: EVA is principally determined by conceptual and US-GAAP related accounting adjustments both affecting net operating profits after taxes, *NOPAT*, and net adjusted capital, *Capital*, and by the weighted average cost of capital, *WACC*.

Year	EBEI	Conceptual adjustments	US-GAAP adjustments	NOPAT	Net assets	Conceptual adjustments	US-GAAP adjustments	Capital	WACC	EVA
1995	698	88	89	875	5,654	-1,140	1,389	5,902	12%	174
1996	802	89	54	945	6,474	-1,326	1,291	6,438	12%	181
1997	713	249	-13	950	6,546	-1,381	1,071	6,236	12%	199
1998	671	305	-157	820	6,046	-1,178	932	5,799	11%	163
1999	724	139	11	874	5,578	-915	908	5,571	11%	263
2000	714	126	-57	782	5,359	-796	865	5,427	10%	260
2001	649	161	57	867	5,562	-845	1,068	5,784	9%	346
2002	525	148	-483	190	5,824	-1,006	1,433	6,251	9%	-341
2003	626	147	105	878	5,963	-1,041	1,640	6,562	8%	322
2004	647	126	81	854	6,440	-984	1,670	7,126	9%	239
2005	707	119	13	839	6,723	-950	1,675	7,448	9%	144
2006	755	115	172	1,042	7,320	-1,074	1,650	7,896	10%	272

Table 15   Campbell Soup Corp.: Economic Value Added (in m USD)

Earnings adjustments increase EBEI on average by 2%, primarily driven by conceptual EVA adjustments. Capital adjustments increase net capital on average by 35%, primarily driven by US-GAAP related EVA adjustments. Over time, EVA of Campbell Soup increases from 174 m USD in 1995 to 272 m USD in 2006, primarily due to the decrease in the cost of capital. EVA is ranging from -341 m USD in 2002 to 346 m USD in 2001, primarily driven by single adjustments for non-capitalized intangibles and pension liabilities (as described later).

The following subsections discuss the calculations of included adjustments and the cost of capital.

---

[455] See section 4.2.5.

Conceptual Adjustments

'Conceptual' EVA adjustments include extraordinary gains and losses, non-operating assets, non-interest-bearing current liabilities, non-operating income and expenses, and financing costs.

Table 16 summarizes the impact of conceptual EVA adjustments for Campbell Soup. Overall, capital adjustments consist primarily of non-interest bearing current liabilities (NIBCLs), while earnings adjustments reflect primarily other financing costs.

Year	Unusual items	Non-operating assets	NIBCLs	Capital adjustments	Non-operating items	Unusual items	Financing costs	EBEI adjustments
1995	424	-299	-1,265	-1,140	-7	0	95	88
1996	419	-414	-1,332	-1,326	-4	-9	102	89
1997	485	-446	-1,420	-1,381	-5	140	115	249
1998	644	-384	-1,439	-1,178	9	169	128	305
1999	746	-381	-1,281	-915	-7	24	122	139
2000	758	-396	-1,159	-796	-7	0	132	126
2001	763	-372	-1,237	-845	4	10	147	161
2002	780	-388	-1,398	-1,006	-1	24	125	148
2003	820	-368	-1,493	-1,041	-4	25	126	147
2004	857	-325	-1,517	-984	-10	18	119	126
2005	865	-275	-1,540	-950	-7	0	126	119
2006	861	-227	-1,708	-1,074	-11	2	124	115

Table 16   Campbell Soup Corp.: Conceptual EVA Adjustments (in m USD)

In the following list, single adjustments are discussed.

- Extraordinary gains and losses: Adjustments for extraordinary items increase earnings and capital of Campbell Soup on average by 33 m USD and 702 m USD, respectively. The earnings adjustment ranges from decreasing earnings by 9 m USD in 1996 to increasing earnings by 169 m USD in 1998. Extraordinary pretax losses occur quite frequently and reflect primarily restructurings, reconfigurations and write-downs of investments. In contrast, extraordinary pretax gains, including a reversal of a prior charge and settlement credits from a class action lawsuit, and after-tax items, from the adoption of new accounting standards, occur more exceptional and do not affect earnings considerably. The cumulative adjustment to capital increases from 424 m USD in 1995 to 861 m USD in 2006. Given that extraordinary losses generally prevail, the capital adjustment is primarily determined by the selection of the initial year to start the accumulation of extraordinary gains and losses.

- Non-operating assets: Campbell Soup carries two types of non-operating assets: construction in progress (CIP) of on average about 200 m USD and other investments of on average about 150 m USD. Overall, capital decreases on average by 356 m USD. According to APB 12, par. 5, CIP reflect capitalized expenditures for property, plant and equipment in the course of construction, which are not available for operating activities in the current period. [456] Other investments represent limited partnership interests (of less than 20% of voting stock) in affordable housing partnership funds, primarily held to gain significant tax credits.

- Non-interest-bearing current liabilities (NIBCLs): Campbell Soup carries average NIBCLs of about 1.4 bn USD, representing on average 23% of total assets. Accounts payable represent on average 41% of NIBCLs, the remainder consists of accrued liabilities, accrued income taxes and dividends payable. NIBCLs decrease capital by on average 23%, therefore, having a much greater impact on capital than adjustments for unusual items and non-operating assets.

- Non-operating income and expenses: Earnings of Campbell Soup are to be reduced on average by 4 m USD, consisting primarily of non-operating interest income. Otherwise, included non-operating pretax charges result from stock price related incentive program expenses; included non-operating after-tax charges represent losses from discontinued operations.

- Financing costs: Financing costs reduce earnings of Campbell Soup on average by 122 m USD, primarily consisting of interest expense. Adjustment for financing costs affect earnings considerably more than adjustments for non-operating and unusual income and expenses.

US-GAAP Related Adjustments

In addition to conceptual EVA adjustments, the study applies various US-GAAP related EVA adjustments, specifically non-capitalized operating leases, historic cost and other reserves, goodwill amortization, non-capitalized intangibles, deferred taxes, pension and postretirement plans, employee stock ownership plan, and stock-based compensation.

The following table summarizes the impact of US-GAAP related adjustments for Campbell Soup. Overall, all capital adjustments increase net capital. On average, 66% of capital adjustments are non-capitalized advertising expenses and 22% of capital adjustments are pension liabilities. Earnings adjustments mostly reduce net earnings, reflecting primarily changes in non-capitalized advertising expenses and deferred taxes.

---

[456] See Stewart (1991, p. 744).

Year	Present value of leases	Reserves	Goodwill amortization	Intangibles	Pension liability	Capital adjustments	Lease interest	Δ Deferred taxes	Δ Reserves	Goodwill amortization	Δ Intangibles	Pension expense adjustment	ESOP contribution	Fair value reserve	EBEI adjustments
1995	0	205	0	812	372	1,389	0	24	-48	0	78	35	0	0	89
1996	0	150	0	883	258	1,291	0	39	-63	0	63	15	0	0	54
1997	0	93	0	948	31	1,071	0	-23	-51	0	66	-5	0	0	-13
1998	0	47	0	947	-62	932	0	-5	-41	0	-68	-28	0	-15	-157
1999	0	31	0	937	-59	908	0	17	9	0	47	-40	0	-22	11
2000	0	36	0	953	-124	865	0	22	1	0	-14	-46	0	-20	-57
2001	0	38	22	979	29	1,068	0	18	4	29	66	-46	0	-14	57
2002	78	43	44	844	424	1,433	10	-115	6	0	-336	-33	0	-15	-483
2003	175	53	44	702	666	1,640	13	57	14	0	52	-7	0	-24	105
2004	189	64	44	724	649	1,670	12	87	8	0	-8	11	0	-29	81
2005	194	61	44	734	643	1,675	13	4	-14	0	28	10	0	-29	13
2006	213	39	44	770	584	1,650	16	126	-30	0	44	16	0	0	172

Table 17   Campbell Soup Corp.: US-GAAP EVA Adjustments (in m USD)

Again, individual adjustments are separately reviewed.

- Non-capitalized operating leases: In 2002, Campbell Soup started disclosing information on minimum rental payments within the note 'Commitments and Contingencies' to the financial statements. Non-capitalized operating leases increase from 75 m USD in 2002 to 213 m USD in 2006. Respective after-tax interest expense at 10% equals on average 13 m USD per annum.

- Historic cost and other reserves: The Campbell Soup Corporation discloses Last-in-First-out (LIFO) inventory reserves within the note 'Inventories' of the annual report. Precautionary deferred tax asset valuation allowances and allowances on receivables are additionally shown within the notes 'Taxes on Earnings' and 'Accounts Receivables', respectively, of the financial statements. The valuation allowance for deferred tax assets is an estimate of the portion of deferred tax assets that is not realizable because of insufficient future taxable income.[457] Allowances for receivables, also referred to as allowances for doubtful accounts, are an estimation made by the firm for receivables that might go uncollected, if losses from uncollectible receivables are probable and estimable.[458] Consequently, both reserves

---

[457] See SFAS No. 109.
[458] See SFAS No. 5, par. 8 and 22.

require some management estimate of unrealizable future income, providing some opportunity for earnings management. All disclosed reserves are reversed as follows: Reserves of on average 71 m USD represent additional capital and annual decreases in reserves of on average 17 m USD reflect a decrease in earnings.

- Goodwill amortization: Campbell Soup only discloses pretax goodwill amortization of 44 m USD in 2001. Consequently, earnings are increased by 29 m USD in 2001 and cumulative goodwill amortization of 44 m USD increases capital in 2001 and succeeding years.

- Non-capitalized intangibles: Campbell Soup Corporation carries non-capitalized intangibles that primarily arise from advertising expenses. Research and development (R&D) expenses, obtained from Worldscope item WC01201, were relatively low, increasing from 54 m USD in 1990 to 99 m USD in 2006. On the contrary, advertising expenses are relatively high, increasing from 1.2 bn USD in 1993 to 1.8 bn USD in 2001 and subsequently decreasing to 1.2 bn USD in 2006. Due to the capitalization of R&D and advertising expenses, capital increases on average by 853 m USD. Changes in net intangibles, affecting earnings, vary from -336 m USD in 2002 to 75 m USD in 1995. Substantial negative performance in fiscal year 2002 is mainly driven by comparably low NOPAT due to a decrease of non-capitalized advertising expenses by 0.3 bn USD. In this year, marketing expenses has been reclassified as a direct reduction of sales. Consequently, the firm disclosed a decrease of marketing and selling expenses by 0.7 bn USD, while marketing expenses were steadily increasing from 1994 to 2001.

- Deferred taxes: The total amount of capital of Campbell Soup is not affected by the reclassification of deferred taxes as equity equivalent; corresponding adjustments to earnings vary from -115 m USD in 2002 to 126 m USD in 2006. While deferred taxes constantly increase from 211 m USD in 1994 to 303 m USD in 2001, there is a substantial decrease in deferred taxes in 2002 to 188 m USD. Notably, considerable negative performance in fiscal year 2002 is also (next to the adjustment for advertising expenses) driven by a reduction in deferred taxes by 115 m USD.

- Pension and postretirement plans: Campbell Soup carries a defined benefit pension plan and a postretirement plan. The projected pension benefit obligation and accumulated postretirement benefit obligation above the fair value of pension plan assets ranges from -124 m USD in 2000 to 666 m USD in 2003. Pension and post-retirement expenses above the services costs vary from -46 m USD in 2001 to 35 m USD in 1995.

- Employee stock ownership plan (ESOP): Campbell Soup Corporation does not carry an ESOP.

- Stock-based compensation: Campbell Soup discloses an average reduction in earnings due to the fair value recognition of stock option grants by 21 m USD for the fiscal years 1998 to 2005.

Weighted Average Cost of Capital

As shown in Table 18, the weighted average cost of capital, *WACC*, is defined based on the after-tax cost of debt, the cost of equity and respective capital weights.

Year	Interest expense	Book value of debt	Effective tax rate	AT cost of debt	Industry beta	Cost of equity	Book value of debt	Market value of equity	Book value of minority interest	Book value of preferred stock	Debt % of capital	Equity % of capital	WACC
1995	123	1,358	32%	6.1%	0.8	13.3%	1,358	5,369	114	0	19.9%	80.1%	11.9%
1996	137	1,666	33%	5.5%	0.8	13.3%	1,666	7,201	90	0	18.6%	81.4%	11.9%
1997	178	2,134	35%	5.4%	0.7	13.0%	2,134	14,825	83	0	12.5%	87.5%	12.0%
1998	194	2,615	36%	4.8%	0.6	12.0%	2,615	24,033	4	0	9.8%	90.2%	11.3%
1999	190	2,944	34%	4.3%	0.6	11.9%	2,944	20,821	0	0	12.4%	87.6%	11.0%
2000	204	3,204	34%	4.2%	0.4	10.8%	3,204	14,365	0	0	18.2%	81.8%	9.6%
2001	222	3,570	34%	4.1%	0.3	10.4%	3,570	12,181	0	0	22.7%	77.3%	9.0%
2002	191	3,847	34%	3.3%	0.3	10.3%	3,847	11,183	0	0	25.6%	74.4%	8.5%
2003	188	3,587	32%	3.6%	0.3	10.3%	3,587	9,479	0	0	27.5%	72.5%	8.5%
2004	177	3,441	32%	3.5%	0.3	10.2%	3,441	10,902	0	0	24.0%	76.0%	8.6%
2005	188	3,173	31%	4.1%	0.4	10.7%	3,173	11,908	0	0	21.0%	79.0%	9.3%
2006	170	3,103	25%	4.1%	0.4	11.1%	3,103	13,278	0	0	18.9%	81.1%	9.8%

Table 18   Campbell Soup Corp.: Weighted Average Cost of Capital (in m USD)

The market cost of debt is approximated by dividing the interest expense on debt by total debt. Campbell Soup shows a cost of debt ranging from 3.3% to 6.1% for fiscal years 1995 to 2006. The cost of debt decreased, since the increase in debt (8% p.a.) outweighed the increase in interest expenses (3% p.a.).

The cost of equity is historically derived based on the 'Capital Asset Pricing Model' (CAPM). While the risk-free rate of 8.6% (derived from 10-year U.S. Treasury bond returns for the years 1992 to 2002) and the market risk premium of 6.0% (derived from Standard & Poor's 500 stock index returns and 20-year U.S. Treasury bond returns for the years 1925 to 1989) apply to all firms and across years, the yearly beta factor of the industry determines the cost of equity of the firm. For Campbell Soup, the cost of equity decreases from 13% in 1995 to 11% in 2006, since the beta factor, which refers to the segment of food producers, decreases from 0.8 to 0.4, respectively.

Regarding the capital weights, market value of equity and book value of minority interest represent on average 81% of capital, while the book value of debt accounts for 19% of capital. The firm has not issued any preferred stock.

Altogether, the WACC decreases from 11.9% in 1995 to 9.8% in 2006, since the cost of debt and equity are both decreasing over time and the capital structure is constant.

Differences between RI and EVA depend on the firm and vary year by year. For example, for Campbell Soup and Alaska Air, RI is consistently higher than EVA, see Figure 27; however, differences between EVA and RI vary considerably for Unifirst.

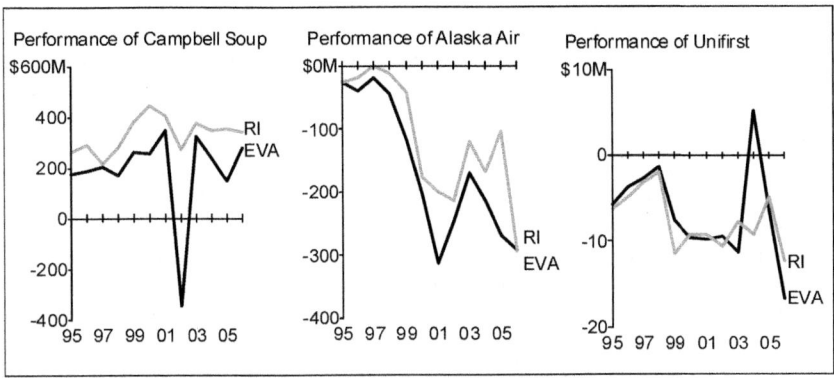

Figure 27   Case Study - Comparison of RI and EVA

The negative EVA of Campbell Soup in 2002 is most prominent, primarily driven by an unexpected decrease in marketing expenses and deferred taxes (both were constantly increasing over 1994 to 2001). Therefore, EVA seems at first sight absurd. However, in fact, it may more correctly reflect investors' impression from the firm's value creation than RI.

Cash Residual Income (CRI)

Next to maintaining a gross asset basis, the CRI metric includes figurative economic depreciation, *ED*, as distinctive component. ED is defined based on based on three components: gross depreciable assets, *GDA*, average asset life, *T*, and weighted cost of capital, *WACC*, as follows:

$$ED_t = GDA_{t-1} \cdot \frac{WACC}{(1+WACC)^T - 1} = GDA_{t-1} \cdot \frac{WACC}{(1+WACC)^{GDA_{t-1}/DA_t} - 1} \cdot {}^{459}$$

Table 19 reviews calculation of economic depreciation for Campbell Soup.

Gross depreciables represent the first key element of economic depreciation, as described by the first shaded section of the table: Gross depreciating assets, being equal to net plants less land and construction in progress (CIP), other intangibles less gross goodwill (or alternatively net goodwill and accumulated goodwill amortization) and accumulated goodwill amortization, and accumulated depreciation and amortization, amounts to about 4 bn USD, on average. Gross depreciable assets are relatively stable over time and consist to 89% of depreciable tangibles.

Highlighted by the second section of the table, the average useful life of assets constitutes the second key component of economic depreciation. Approximated by dividing gross depreciables by depreciation and amortization expenses, the average useful life of the assets is on average about 14 years. Due to year-to-year fluctuations, it ranges between 12.8 and 17.5 years. There is no clear decrease of the assets' life over the analyzed eleven-year period, suggesting that Campbell Soup regularly replaces obsolete assets with new assets of equivalent useful lives.

WACC estimates, representing the third basic element of figurative depreciation, are obtained from EVA calculations.

Based on these three elements, economic depreciation is on average 137 m USD (see last row of the table). Accounting for the time value of money, it is considerably lower than book depreciation and amortization of on average 281 m USD. Economic depreciation ranges from 93 m USD to 188 m USD, due to the following: In 1998, a 33% increase in net plants resulted in a considerable increase in the average asset life to 17.5 years. Therefore, economic depreciation decreases to the minimum value of 93 m USD in 1998. In 2002, Campbell Soup started disclosing the amount of other non-amortizable intangible assets, decreasing the amount of depreciable assets substantially. Consequently, also the approximated asset life decreases to 12.8 years and economic depreciation increases to the maximum value of 188 m USD in 2002.

---

[459] See section 4.2.5.

Year	Net plant	Land	Construction in progress	Accumulated depreciation	Other intangibles	Gross goodwill	Net goodwill	Other non-amortizables	Accumulated amortization	Accum. goodwill amortization	Gross depreciating assets	Depreciation & amortization	Asset life	WACC	ED
1995	2,584	101	237	1,670	1,715	1,334			101	0	4,398	294	12.5	12%	142
1996	2,681	99	332	1,809	1,808	1,407			131	0	4,591	326	13.5	12%	147
1997	2,560	85	214	1,867	1,793	1,478			133	0	4,576	328	14.0	12%	141
1998	1,723	54	166	1,437	1,904	1,667			179	0	3,356	261	17.5	11%	93
1999	1,726	50	184	1,491	1,910	1,697			220	0	3,416	255	13.2	11%	125
2000	1,644	43	162	1,652	1,767	1,570			263	0	3,551	251	13.6	10%	132
2001	1,637	50	133	1,740	2,451	1,856			306	22	4,095	266	13.3	9%	148
2002	1,684	53	230	1,949	2,534	1,881	1,581	939	387	44	3,451	319	12.8	9%	188
2003	1,843	66	145	2,169	2,821		1,803	1,005	390	44	4,160	243	14.2	8%	134
2004	1,901	70	192	2347	2,995		1,900	1,082	391	44	4,346	260	16.0	9%	130
2005	1,987	69	208	2,524	3,009		1,950	1,047	392	44	4,594	279	15.6	9%	135
2006	1,954	56	245	2,543	2,361		1,765	588	388	44	4,548	289	15.9	10%	132

Table 19   Campbell Soup Corp.: Economic Depreciation (in m USD)

Overall, CRI differs from RI by the replacement of book depreciation by lower economic depreciation and an additional capital charge on accumulated amortization and depreciation. Capital charges are based on the WACC estimate (consistently obtained from EVA calculations) and gross capital (average interest-bearing capital before accumulated depreciation and amortization). Then, CRI is defined as:[460]

$$CRI_t = \left(EBEI_t + IntExp_t \cdot (1-t) + DA_t\right) - ED_t - WACC_t \cdot \left(\frac{NA_{t-1} + NA_t}{2} + \frac{ADA_{t-1} + ADA_t}{2} \right)$$

Campbell Soup has an average CRI of 260 m USD, see results in Table 20.

Year	EBEI before interest	Depreication & amortization	Gross earnings	Economic depreciation (ED)	Capital excluding NIBCLs	Accumulated depreciation	Accumulated amortization	Gross capital	WACC	CRI
1995	781	294	1,075	142	4,389	1,559	96	6,043	12%	215
1996	894	326	1,220	159	5,142	1,740	116	6,998	12%	231
1997	828	328	1,156	141	5,126	1,838	132	7,096	12%	160
1998	796	261	1,057	101	4,608	1,652	156	6,416	11%	229
1999	849	255	1,104	125	4,297	1,464	200	5,961	11%	325
2000	849	251	1,100	132	4,200	1,572	242	6,013	10%	389
2001	795	266	1,061	108	4,325	1,696	285	6,306	9%	386
2002	651	319	970	188	4,426	1,845	347	6,617	9%	219
2003	753	243	996	134	4,47	2,059	389	6,918	8%	276
2004	768	260	1,028	130	4,924	2,258	391	7,572	9%	245
2005	836	279	1,115	135	5,183	2,436	392	8,010	9%	233
2006	883	289	1,172	132	5,612	2,534	390	8,535	10%	207

Table 20   Campbell Soup Corp.: Cash Residual Income (in m USD)

Figure 28 compares CRI with RI for Campbell Soup, Alaska Air, and Unifirst. Suppositions on lower volatility of CRI versus RI due to reduced built-in performance improvements cannot be tested, since all three firms replace the asset portfolio continuously, maintaining approximately the same average asset age over time.

However, the figure shows that average levels of RI and CRI differ, as suggested by theory: In the early (or last) years of the asset life, RI understates (overstates) performance; therefore, CRI shall be higher (lower) than RI. [461] Consistent to theoretical assumptions, Alaska Air, carrying a quite 'young' asset portfolio (titled 'ambitious investor'), shows CRI levels consistently above RI levels. Unifirst, owning

---

[460] See section 4.2.5.
[461] See section 3.1.

fairly 'old' assets (titled 'mature investor'), shows CRI levels consistently below RI levels. However, differences are much smaller for Alaska Air than for Unifirst, since CRI (referring to gross assets) is generally lower than RI (referring to net assets). Therefore, slightly higher values of CRI with respect to RI, as shown in the figure below, imply a considerably better performance measured by CRI with reference to the outcome measured by the RI metric. For the same reasons, Campbell Soup that carries a portfolio of 'medium-aged' assets shows CRI levels systematically below RI levels.

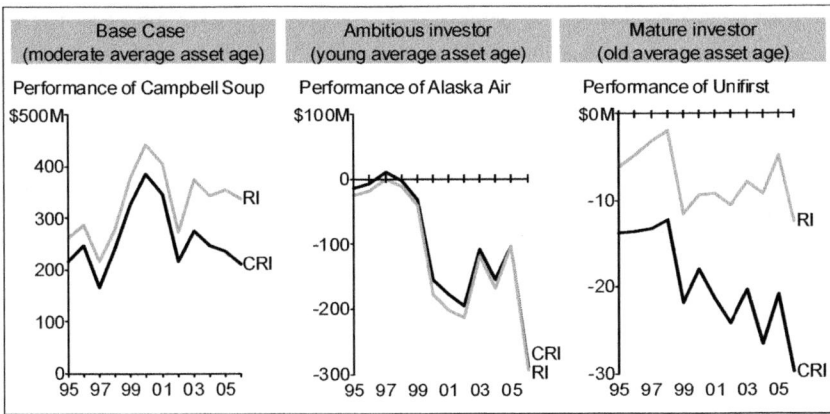

Figure 28   Case Study - Comparison of RI and CRI

Next, as shown in Figure 29, corresponding profitability measures of RI and CRI are compared; notably, as already discussed, both RONA and ROGA refer to the same reference value (WACC).[462] ROGA overcomes deficits of RONA in determining the level of shareholder value creation, namely underestimation of the firm's value creation in early years and overestimation in late years of the asset's useful life. Furthermore, ROGA referring to gross assets can be seen as a 'yearly component' of the average IRR of the company historical investments.

Campbell Soup shows an average RONA of 17% and an average ROGA of 14% (both higher than the average 10% cost of capital in the Food Producers' segment), indicating that consistently value was added to net and gross assets.

Alaska Air, which invested heavily in new assets, shows an average RONA of 4% and an average ROGA of 6% (both below the 9% average cost of capital in the 'Travel & Leisure' segment). This means that Alaska Air destroyed value with respect to its net and gross assets. However, the depreciation-adjusted ROGA measure is higher than RONA, although ROGA is expected to be lower than RONA (since referring to

---

[462] See section 3.3.1.

gross instead of net assets).Therefore, ROGA primarily avoids underestimations of profitability due to book depreciation.

Unifirst, which primarily retained old assets, shows an average RONA of 9% and an average ROGA of 7% (both below the 11% average cost of capital in the 'Personal Goods' segment), indicating that also Unifirst destroyed shareholder value. ROGA levels below RONA levels are consistent with theory, asserting that ROGA avoids overestimation of performance given relatively 'old' investments.

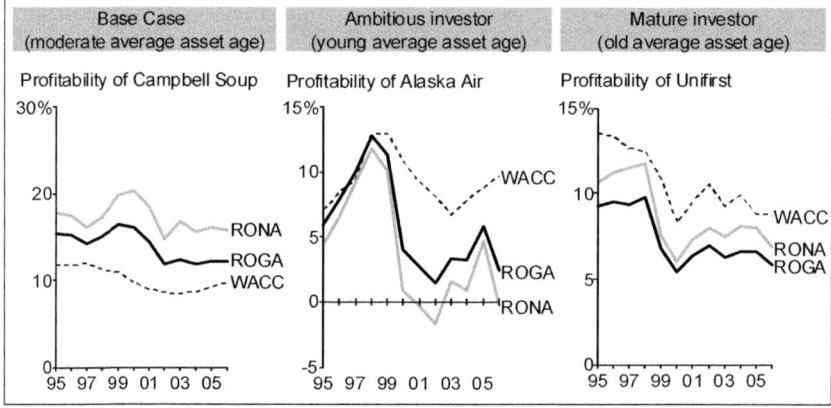

Figure 29   Case Study - Comparison of RONA and ROGA

Cash Economic Value Added (CEVA)

Finally, CEVA is defined based on its four key elements: adjusted gross earnings, economic depreciation, ED, adjusted gross capital, and the weighted average cost of capital, $WACC$, as: [463]

$$CEVA_t = \left(NOPAT_t + DA_t - GA_t\right) - ED_t - WACC_t \cdot$$
$$\cdot \left(Capital_{t-1} + Capital_t \Big/ 2 + ADA_{t-1} + ADA_t \Big/ 2 - AGA_{t-1} + AGA_t \Big/ 2\right).$$

Table 21 outlines the CEVA calculation for Campbell Soup.

Year	Adjusted net earnings	Depreciation & amortization	Goodwill amortization	Adj. gross profit	ED	Adjusted net assets	Accumulated depreciation	Accumulated amortization	Accum. goodwill amortization	Adj. gross capital	WACC	CEVA
1995	875	294	0	1,169	142	5,902	1,559	96	0	7,556	12%	130
1996	945	326	0	1,271	147	6,438	1,740	116	0	8,294	12%	139
1997	950	328	0	1,278	141	6,236	1,838	132	0	8,206	12%	148
1998	820	261	0	1,081	93	5,799	1,652	156	0	7,607	11%	126
1999	874	255	0	1,129	125	5,571	1,464	200	0	7,234	11%	210
2000	782	251	0	1,033	132	5,427	1,572	242	0	7,240	10%	204
2001	867	266	-44	1,089	148	5,784	1,696	285	-22	7,742	9%	244
2002	190	319	0	509	188	6,251	1,845	347	-44	8,398	9%	-393
2003	878	243	0	1,121	134	6,562	2,059	389	-44	8,965	8%	226
2004	854	260	0	1,114	130	7,126	2,258	391	-44	9,731	9%	145
2005	839	279	0	1,118	135	7,448	2,436	392	-44	10,231	9%	29
2006	1,042	289	0	1,331	132	7,896	2,534	390	-44	10,775	10%	148

Table 21   Campbell Soup Corp.: Cash Economic Value Added (in m USD)

CEVA represents a modification of EVA that characteristically adjusts for accounting depreciation. Contrary to the depreciation adjustment included in the CRI metric, the adjustment within CEVA excludes goodwill amortization (since the EVA method yet adjusts for goodwill). Since economic depreciation is independent from the adjustment goodwill, CRI and CEVA differ by including or excluding goodwill amortization within gross earnings and gross capital, respectively. The table highlights

---

[463] See section 4.2.5.

the separate handling of goodwill amortization in the course of the calculation of CEVA which ranges from -393 m USD in 2002 to 244 m USD in 2001.

Overall, EVA and CEVA, both being adjust for accounting-related distortions, show a similar pattern of value creation, see Figure 30.

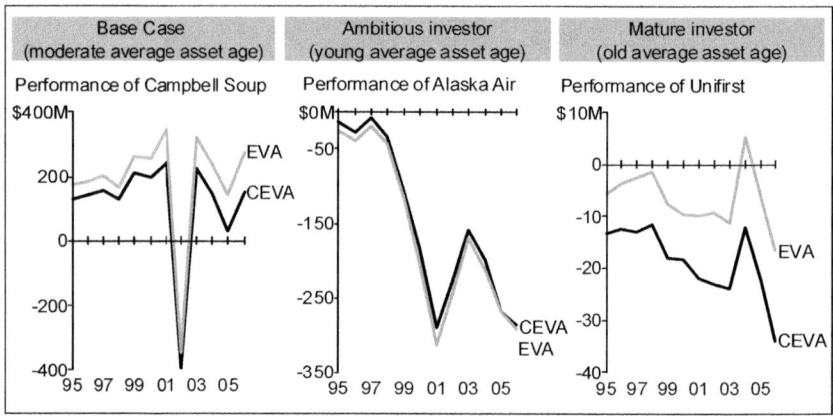

Figure 30   Case Study - Comparison of EVA and CEVA

Yet mentioned above, negative value creation of Campbell Soup in 2002 corresponds to a decrease in marketing costs and deferred taxes. Similarly, value added of Unifirst is relatively high in 2004 due to a substantial decrease in deferred taxes by 17 m USD. In 2001, indicated value creation of Alaska Air is negative, due to a substantial extraordinary gain of 49 m USD that improves accounting earnings but is not expected to recur.

Average EVA levels are generally lower than average CEVA levels for firms with relatively young assets, and vice versa. Again, differences are much smaller for Alaska Air than for Unifirst, since CEVA (which refers to gross assets) is generally lower than EVA (which refers to net assets). And, both measures are in most cases consistent with respect to whether the firm has created or destroyed value.

As shown in Figure 31, also the levels of the two corresponding profitability measures, ROCE and CROCE, differ consistently. Since ROCE and CROCE refer to a different capital basis (net and gross assets, respectively), CROCE is supposed to be generally lower than ROCE. However, for Alaska Air, an average CROCE level above the average ROCE level unveils that ROCE understates value creation if assets are relatively young; similarly, for Unifirst, CROCE being lower than ROCE shows that ROCE overstates value creation for relatively old assets.

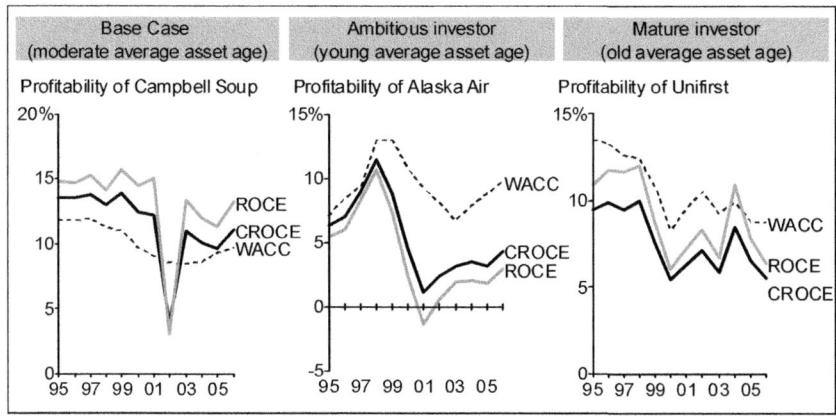

Figure 31   Case Study - Comparison of ROCE and CROCE

5.1.2   Market-Based Measures of Shareholder Value Creation

This study defines two market measures that determine shareholder wealth: The market value of the firm and abnormal returns of the firm's equity.[464]

Market Value of the Firm (MV)

'Market Value' (MV), defined as $MV_{t,t-1} = MVE_{t,t-1} + NA_{t,t-1} - CE_{t,t-1}$,[465] consists of the market value of equity and the book value of capital other than equity (such as debt and preferred stock). As shown in the following table, MV of Campbell Soup increases from 8.4 bn USD in 1995 to 29.0 bn USD in 1998, then, decreasing to 15.0 bn USD by 2003 and, again, increasing modestly to 18.4 bn USD by 2006.

Changes in MV are primarily determined by changes in the market value of equity (see first row of the table), since market value of equity accounts to 72% of firm value. While total assets (see second row) increase modestly at 2% p.a. from 5.7bn USD to 7.3 bn USD in 2006, they do not affect changes in MV. Book value of equity (see third row of the table) is comparably low, on average 1.0 bn USD. Notably, the firm discloses shareholders' deficits for fiscal years 2001 and 2002 due to investments in own stock. Treasury stock increased from 550 m USD in 1995 to 4.9 bn USD in 2001.

---

[464] See section 4.2.5.
[465] See section 4.2.5.

Year	Market value of equity	Total net assets	Book value of equity	Market value of the firm (MV)
1995	5,185	5,654	2,468	8,370
1996	6,779	6,474	2,742	10,510
1997	13,728	6,546	1,420	18,854
1998	23,877	6,046	874	29,049
1999	21,190	5,578	235	26,533
2000	15,668	5,359	137	20,890
2001	12,226	5,562	-247	18,034
2002	11,449	5,824	-114	17,387
2003	9,423	5,963	387	14,999
2004	10,848	6,440	874	16,414
2005	11,752	6,723	1,270	17,205
2006	12,803	7,320	1,768	18,355

Table 22   Campbell Soup Corp.: Market Value of the Firm (in m USD)

Abnormal Stock Returns (AR)

'Abnormal Returns' (AR) are defined as common stock return less MSCI return, notationally $AR_{i,t} = \Pi_{\tau=1}^{12}(1 + r_{i,\tau}) - \Pi_{\tau=1}^{12}(1 + r_{m,\tau}) - 1$.[466] As shown in the following table, AR of Campbell Soup are quite volatile ($SD = 28\%$), but balance out over time ($M = -0.3\%$). For fiscal years 1996 and 1999 to 2001, differences in performance between the equity and the MSCI index are quite substantial, ranging from -53% to 41%. Sizeable differences are due to numerous firm- and industry-specific factors, too complex to analyze and describe here.

Year	Stock return	Morgan Stanley Capital Index (MSCI) return	Abnormal returns (AR)
1995	-50%	-68%	18%
1996	59%	20%	39%
1997	35%	34%	1%
1998	10%	17%	-7%
1999	-14%	26%	-40%
2000	-41%	11%	-53%
2001	12%	-29%	41%
2002	-19%	-18%	-1%
2003	21%	34%	-12%
2004	6%	10%	-4%
2005	13%	13%	0%
2006	25%	10%	15%

Table 23   Campbell Soup Corp.: Abnormal Stock Returns (in m USD)

---

[466] See section 4.2.5.

5.1.3   Examination of Interrelationships

Summarizing accounting and stock performance, Figure 32 shows firm performance quantified via six accounting metrics (CFO, EBEI, RI, CRI, EVA, and CEVA) in the first row and shareholder wealth defined via two market measures (AR and MV) in the second row. The arrows pointing from accounting performance towards market performance stands for the assumption that accounted performance explains (to some extent) stock prices.

As shown in the upper section of Figure 32, accounting metrics show fairly consistent relative performance levels and volatility across firms: The traditional performance measures, CFO and EBEI, show highest levels of firm performance, since they avoid charges for capital sources such as preferred and common stock, which value-based measures include. CFO is always higher then EBEI, pointing towards the predominance of non-cash charges, such as expenses for restructuring and depreciation and amortization. EBEI is generally less volatile than CFO due to smoothing accruals. Changes in depreciation-adjusted CRI (CEVA) are similar to changes in unadjusted RI (EVA). However, CRI and CEVA levels are generally higher than the underlying RI and EVA levels, respectively, if the average age of the firm is relatively young, and vice versa.

The lower section of Figure 32 summarizes selected market-based performance measures for Campbell Soup, Alaska Air, and Unifirst: Market values generally increase over the period of the empirical study. Market value of Campbell Soup and Unifirst approximately doubled and market value of Alaska Air increased about two times. Generally, increases (decreases) in market value suggest additions to (subtractions from) shareholder value. However, this approach neglects that dividends and other stock adjustments additionally add to shareholder value. Contrary to market value that exclusively refers to the firm performance, excess returns provide statements on the firm's market performance relative to overall market dynamics: - 0.3% average abnormal returns of Campbell Soup imply that the firm created (or destroyed) value in line with the overall value creation (or destruction) of the market. 6.9% and 4.2% average abnormal returns of Alaska Air and Unifirst, respectively, mean that those firms outperformed the market.

Correlations between accounting measures and market value or abnormal returns differ substantially by metric and firm. For Campbell Soup, CRI ($r_S = 0.29$) and CFO ($r_S = 0.24$) are most highly correlated to market value and abnormal returns, respectively. For Alaska Air, CEVA is most highly correlated with market value ($r_S = -0.88$), while CFO is most highly correlated with abnormal returns ($r_S = -0.33$). However, negative correlations of CFO with excess returns contradict intuition. For Unifirst, EBEI ($r_S = 0.80$) and CRI ($r_S = 0.31$) show greatest correlations with firm value and excess returns, respectively. In this case, positive correlations of permanently negative CRI with variable abnormal returns also interfere with intuition.

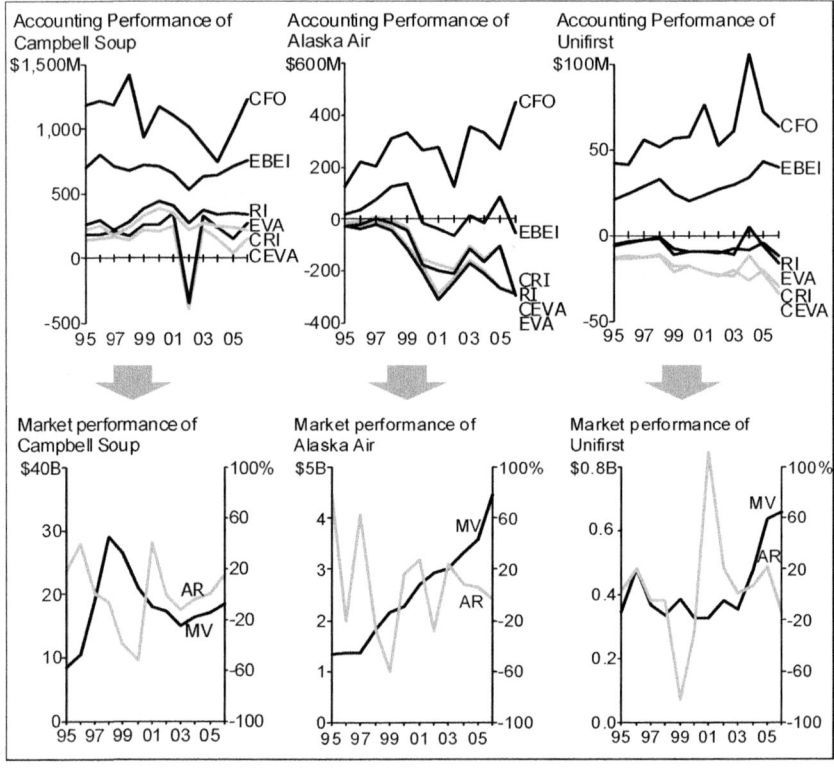

Figure 32   Case Study - Relation of Accounting Metrics to Market Performance

5.2 Regression Analysis

Based on the full sample of 2,147 firm-years, regression results vary considerably.[467] Therefore, the following results are derived from a sample that excludes extreme values above and below eight standard deviations from the median. Results refer to panel data covering 201 U.S. firms over the time period from 1995 to 2006.

Results by industry or fiscal year are not shown, since results on subgroups would suffer from an insufficient sample size. Moreover, the sample is not partitioned into adopters of a VBM system versus non-adopters, since classification requires to some extent subjective judgments on the application of a value-based system.

5.2.1 Descriptive and Correlations Statistics

Based on 2,136 firm-year observations, the following table provides descriptive statistics and correlations of firm-year observations used to test the relationship between accounting measures and the market value of the firm.

	Dependent variable	Independent variables (all divided by WACC and BV)					
Descriptives	MV / BV	CFO	EBEI	RI	CRI	EVA	CEVA
M	1.781	1.061	0.594	-0.093	-0.181	-0.191	-0.282
Median	1.412	1.045	0.596	-0.111	-0.192	-0.204	-0.290
SD	1.203	0.792	0.596	0.579	0.584	0.627	0.651
Correlations							
MV / BV	1.000						
CFO	0.301**	1.000					
EBEI	0.433**	0.522**	1.000				
RI	0.440**	0.480**	0.922**	1.000			
CRI	0.407**	0.458**	0.910**	0.973**	1.000		
EVA	0.380**	0.447**	0.670**	0.663**	0.674**	1.000	
CEVA	0.340**	0.415**	0.644**	0.625**	0.684**	0.977**	1.000

** Significance at a 1% level

Table 24 Descriptive and Correlation Statistics Based on Firm Value

Market value of the firm, *MV*, (described in the first column of Table 24) equals on average 1.78 times the book value of its capital. The multiple of 1.78 is relatively low, since the measure is defined to recognize only the market value of common equity, which on average just accounts for 42% of capital.

---

[467] Results based on the full sample, before eliminations of extreme values, are provided in the appendix, Table 37, Table 38, Table 39, and Table 40.

The following columns of the table describe statistics on accounting measures, capitalized by the cost of capital and scaled by book value.

Consistent to results from the case study, CFO (divided by the cost of capital, $WACC$, and the book value of capital, BV) shows the largest mean ($M = 1.1$), followed by EBEI ($M = 0.6$), RI ($M = -0.1$), CRI and EVA ($M = -0.2$), and CEVA ($M = -0.3$). CFO with the largest mean implies that earnings' accruals are for the most part negative, consistent to common non-cash expenses, such as depreciation and amortization. Residual income measures RI, CRI, EVA and CEVA are on average slightly negative, that means that the sample of firms considered in this study shows difficulties to earn in the long-term above capital costs, while imposing suspicion on a potential over-estimation of the cost of equity.

Moreover, the earnings measure EBEI and the two residual income measures RI and CRI, which both primarily represent earnings, have the lowest standard deviation (SD = 0.6) among accounting measures. Low volatility of earnings measures is consistent with the smoothing effects of accruals.

Finally, contemporaneous correlations of accounting measures with market value, MV, are all positive and significant at the 1% level. Correlations are highest for EBEI and RI ($r_S = 0.4$), consistent with investors' focus on earnings.[468]

Next, Table 25 provides descriptive statistics and correlations of firm-year observations used to examine the relationship between accounting measures and stock returns, based on 1,925 firm-year observations.

	Dependent variable	Independent variables (all divided by MVE)					
Descriptives	AR	CFO	EBEI	RI	CRI	EVA	CEVA
M	0.040	0.167	0.078	-0.040	-0.054	-0.058	-0.072
Median	-0.009	0.121	0.066	-0.014	-0.025	-0.026	-0.040
SD	0.436	0.214	0.137	0.142	0.147	0.169	0.177
Correlations							
MV / BV	1.000						
CFO	0.270**	1.000					
EBEI	0.124**	0.346**	1.000				
RI	-0.011	0.059**	0.760**	1.000			
CRI	-0.042	0.032	0.743**	0.982**	1.000		
EVA	-0.004	0.018	0.410**	0.603**	0.619**	1.000	
CEVA	-0.029	-0.008	0.401**	0.594**	0.640**	0.986**	1.000

** Significance at a 1% level

Table 25   Descriptive and Correlation Statistics Based on Stock Returns

---

[468] See (Foster, 1986; Lev, 1989)

Although the average annual abnormal return is 4.0%, the median return of -0.9% implies that the sample is not biased towards over- or underperforming firms, as shown in the first column of the table above.

Accounting measures (described in the following columns) are yet scaled by the market value of equity. Consistent to the previous data set, CFO ($M = 0.2$) has the largest mean, followed by EBEI, and residual income measures are on average negative. Again, earnings-based measures EBEI, RI, and CRI ($SD = 0.1$) have the lowest standard deviations. While correlations of cash flow ($r_s = 0.3$) and earnings ($r_s = 0.1$) with abnormal returns are positive and significant at the 1% level, residual income measures exhibit slightly negative and insignificant correlations with returns, questioning the validity of the expectancy model.

5.2.2 Relative Information Content: Apparent Predominance of Earnings

The regression analysis assumes that accounting measures determine firm values and stock returns. To examine relative information content, the levels of adjusted R^2s from six separate regressions, i.e., one for every metric, are compared.

Regression Analysis on Market Value

According to the first research hypothesis, competing performance metrics are evaluated with respect to their explanation of firm values.[469]

Table 26 shows adjusted R^2s from the regressions for each metric, based on 2,136 firm-year observations.

Rank	(1)	>	(2)	>	(3)	>	(4)	>	(5)	>	(6)
Measure	RI		EBEI		CRI		EVA		CEVA		CFO
Adj. R^2	0.194		0.187		0.165		0.144		0.115		0.090
p-value		(0.000)		(0.015)		(0.000)		(1.000)		(0.000)	
			(1.000)		(0.000)		(0.000)		(0.000)		
					(0.000)		(0.000)		(0.000)		
					(0.000)		(0.000)				
							(0.000)				

Table 26   Relative Information Content Tests from Associations with Firm Value

Performance metrics are listed in descending order of adjusted R^2s: The metric with the highest adjusted R^2 is shown left and the one with the lowest adjusted R^2 is shown right. p-values from one-tailed Cox tests of relative information content are shown centered in parentheses for each of the 15 comparisons of adjusted R^2 levels. Underlying is the non-nested test of the null hypothesis of difference between the

---

[469] The underlying basic regression equation is $\dfrac{MV_t}{BV_{t-1}} = \beta_0 + \beta_1 \cdot \dfrac{X_t / WACC_t}{BV_{t-1}} + \varepsilon_t$, where $X =$ performance measures CFO, EBEI, RI, CRI, EVA, and CEVA, see section 4.2.2.

larger adjusted R^2 versus the smaller adjusted R^2. E.g., the first row leftmost p-value refers to a comparison between the first and second ranked metric. The second row leftmost p-value represents a comparison between the first and third ranked metric. The last row p-value represents a comparison between the first and third ranked metric, etcetera. Due to the formulation of the null hypothesis, ordinary significance levels do not apply. In this study, a difference in adjusted R^2s is considered significant if the p-value is 0.100 or greater.

The table outlines that EBEI and RI are most value-relevant (adj. $R^2 = 19\%$), followed by CRI (adj. $R^2 = 17\%$), EVA (adj. $R^2 = 14\%$), CEVA (adj. $R^2 = 12\%$), and CFO (adj. $R^2 = 9\%$). While metrics show differing information content, differences are mostly not significant (p-value = 0.000). However, net measures RI and EVA contain significant information advantage over cash-based gross measures CRI and CEVA, respectively (p-value = 1.000). To summarize, net earnings measures RI and EBEI are superior in explaining firm value, value-relevance of depreciation-adjusted CRI is comparable to those of leading measures, and CFO is least value relevant.

Overall results are consistent with previous studies. E.g., Biddle et al. (1997) and Kramer and Pushner (1997) provided similar evidence for the outperformance of earnings in explaining market values. Furthermore, O'Byrne (1996), Biddle et al. (1997), Feltham et al. (2004), and West and Worthington (2004) provided consistent evidence on the low information content of cash flows. However, latest evidence on the outperformance of EVA, as indicated by Feltham et al. (2004) and West and Worthington (2004), cannot be reconfirmed.

Additionally, results suggest that using EVA data published by Stern Stewart overestimates the value-relevance of EVA. At first, O'Byrne (1996) presented a considerably high association level of EVA (adj. $R^2 = 31\%$) from U.S. ranking data published by the consultancy Stern Stewart & Company. Then, Tsuji (2006) finds considerably lower value-relevance, based on independently estimated EVA data for Japanese firms (adj. $R^2 = 2\%$). This study suggests that value-relevance of individual EVA data (adj. $R^2 = 14\%$) is higher than the information content measured by Tsuji. Differences to Tsuji may result from differing accounting principles and market dynamics for Japanese firms and a different selection and definition of adjustments when estimating EVA data.[470]

Four model extensions examine the sensitivity of the results, as shown in Table 27.

---

[470] Used adjustments include primarily provisions, minority interest, and consolidated adjustment accounts.

Rank	(1)	>	(2)	>	(3)	>	(4)	>	(5)	>	(6)
A: Sign-based coefficients											
Measure	EBEI		RI		CRI		EVA		CEVA		CFO
Adj. R^2	0.279		0.275		0.262		0.184		0.161		0.112
B_1	1.376		1.749		1.910		1.324		1.359		0.595
B_2	-0.595		0.145		0.128		0.260		0.223		-0.264
p-value		(0.000)		(0.996)		(0.000)		(1.000)		(0.000)	
				(0.000)		(0.000)		(0.000)		(0.000)	
						(0.000)		(0.000)		(0.000)	
								(0.000)		(0.000)	
										(0.000)	
B: Non-capitalized performance											
Measure	EBEI		RI		CRI		CFO		EVA		CEVA
Adj. R^2	0.239		0.189		0.164		0.148		0.134		0.104
p-value		(0.436)		(1.000)		(0.000)		(0.000)		(1.000)	
				(0.997)		(0.000)		(0.000)		(0.000)	
						(0.000)		(0.000)		(0.000)	
								(0.000)		(0.000)	
										(0.000)	
C: Two-year market values											
Measure	RI		EBEI		CRI		EVA		CEVA		CFO
Adj. R^2	0.206		0.190		0.177		0.143		0.115		0.094
p-value		(0.003)		(0.000)		(0.000)		(1.000)		(0.000)	
				(1.000)		(0.000)		(0.000)		(0.000)	
						(0.000)		(0.000)		(0.000)	
								(0.000)		(0.000)	
										(0.000)	
D: Five-year periods											
Measure	EBEI		RI		EVA		CRI		CEVA		CFO
Adj. R^2	0.103		0.064		0.063		0.046		0.044		0.027
p-value		(1.000)		(0.000)		(0.000)		(0.000)		(0.000)	
				(0.000)		(1.000)		(1.000)		(0.000)	
						(1.000)		(0.000)		(0.000)	
								(0.003)		(0.000)	
										(0.100)	

Table 27   Sensitivity Tests on Associations of Metrics with Firm Value

Previous regressions constrained the coefficients to be equal for positive and negative performance. To examine suppositions that the coefficients depend on the sign of performance, Panel A shows results from sign-based coefficients.[471] Adjusted R^2s increase considerably, while relative rankings largely remain: EBEI (adj. R^2 = 28%) dominates, followed by RI (adj. R^2 = 28%), CRI (adj. R^2 = 26%), EVA (adj. R^2 = 18%), CEVA (adj. R^2 = 16%), and CFO (adj. R^2 = 11%). As under constrained coefficients, differences in information content are generally insignificant (p-value = 0.000), except information advantages of RI over CRI (p-value = 0.996) and of EVA over CEVA (p-value = 1.000). Consequently, this supplemental analysis supports basis results. Consistent with prior studies, coefficients of positive performance are generally larger than those of negative performance.[472]

So far, performance has always been capitalized by the cost of capital, following valuation theory. Contrarily, a regression model may alternatively avoid capitalization of performance to reduce errors from perpetuity assumptions.[473] Results, as provided in Panel B, show that value-relevance increases only for traditional measures, which valuation theory does not directly support: While EBEI (adj. R^2 = 24%) is still leading in value-relevance, CFO (adj. R^2 = 15%) has now more information content than EVA and CEVA. As before, net measures RI and EVA significantly outperform the depreciation-adjusted equivalents CRI and CEVA, respectively (p-value = 1.000). However, EBEI gained significant information advantage versus RI (p-value = 0.436) and CRI (p-value = 0.997). Likewise, Biddle et al. (1997) found evidence for the significant outperformance of EBEI. Therefore, this modification further underlines the dominance of earnings.

Next, current-year market value is replaced with average market value over the contemporaneous and the subsequent fiscal year,[474] allowing sufficient time for stock prices to impound accounting information. As shown in Panel C, the modification does not affect basic results. RI (adj. R^2 = 21%) and EBEI (adj. R^2 = 19%) are still leading. Furthermore, results from significance tests are unaffected. Therefore, current-year investor communication and interim results seem to affect market value more than lagged information from annual reports.

Finally, analysis of contemporaneous 5-year changes of performance and firm value addresses Stewart's claim that EVA shows the maximum information content on five-

---

[471] The underlying modified regression equation is

$$\frac{MV_t}{BV_{t-1}} = \beta_0 + \beta_1 \cdot \frac{X_t^{pos}/WACC_t}{BV_{t-1}} + \beta_2 \cdot \frac{X_t^{neg}/WACC_t}{BV_{t-1}} + \varepsilon_t, \text{ see section 4.2.2.}$$

[472] Yet several studies provided evidence for earnings (Burgstahler & Dichev, 1997; Collins et al., 1999; Hayn, 1995) and residual income (Giner & Iniguez, 2006).

[473] The underlying modified regression equation is $\frac{MV_t}{BV_{t-1}} = \beta_0 + \beta_1 \cdot \frac{X_t}{BV_{t-1}} + \varepsilon_t$, see section 4.2.2.

[474] The underlying modified regression equation is $\left(\frac{MV_t}{BV_{t-1}} + \frac{MV_{t+1}}{BV_t} \right)/2 = \beta_0 + \beta_1 \cdot \frac{X_t/WACC_t}{BV_{t-1}} + \varepsilon_t$,

see section 4.2.2.

year data.[475] As shown in Panel D, the relative ranking holds while overall association levels decrease considerably: EBEI (adj. R^2 = 10%) is most value-relevant, followed by RI and EVA (adj. R^2 = 6%), CRI (adj. R^2 = 5%), CEVA (adj. R^2 = 4%), and CFO (adj. R^2 = 3%). However, not only RI and EVA significantly outperform gross equivalents, but also EBEI significantly dominates RI (p-value = 1.000), CRI (p-value = 1.000), and CFO (p-value = 0.100). Value-relevance of EBEI is in line with prior findings by Tsuji (2006). Overall, results do not provide evidence of suppositions. Altogether, the analysis of model variations reconfirms basic results.

So far, adjusted R^2s from the basic regression model, ranging from 9% to 19%, are relatively low. Introducing different regressions for positive and negative performance yet improves regression results considerably, showing adjusted R^2s ranging from 11% to 28%. Following O'Byrne (1996), additional size and industry variables may further improve the regression results. In fact, additional analysis has shown that adjusted R^2s increase, while the ranking of competing measures largely prevails.[476]

Regression Analysis on Stock Returns

According to the third research hypothesis, level and changes of firm performance are related to abnormal returns.[477]

Table 28 summarizes results from 1,925 firm-year observations.

Rank	(1)	>	(2)	>	(3)	>	(4)	>	(5)	>	(6)
Measure	EBEI		RI		CFO		CRI		CEVA		EVA
Adj. R^2	0.075		0.074		0.073		0.063		0.004		0.004
p-value		(0.000)		(0.000)		(0.000)		(1.000)		(0.894)	
			(0.000)		(1.000)		(0.000)		(1.000)		
					(0.783)		(0.159)		(0.000)		
							(0.000)		(0.303)		
							(0.000)				

Table 28   Relative Information Content Tests referring to Stock Returns

---

[475] As shown in section 4.2.2, the underlying modified regression equation is

$$\frac{MV_{t+5}}{BV_{t+4}} - \frac{MV_t}{BV_{t-1}} = \beta_0 + \beta_1 \cdot \left(\frac{X_{t+5}/WACC_{t+5}}{BV_{t+4}} - \frac{X_t/WACC_t}{BV_{t-1}} \right) + \varepsilon_t .$$

[476] E.g., an analysis of the selected 201 firms (after excluding extreme values above one standard deviation and winsorizing data to 4 standard deviations) shows: Adjusted R^2s increase on average by 1% due to the addition of the size variable, but by about 5% due to the addition of industry dummy variables. However, for instance, earnings dominates other measures regardless of the model specification.

[477] The underlying basic regression equation (see section 4.2.2) is:

$$AR_t = \beta_0 + \beta_1 \cdot \frac{X_t}{MVE_{t-1}} + \beta_2 \cdot \frac{(X_t - X_{t-1})}{MVE_{t-1}} + \varepsilon_t .$$

EBEI (adj. R^2 = 8%), RI and CFO (adj. R^2 = 7%), and CRI (adj. R^2 = 6%) are leading with similar information content, while value-relevance of CEVA and EVA (adj. R^2 = 0.4%) is particularly low. Information advantage of basic residual income measures is significant: RI significantly outperforms CRI (p-value = 1.000), CEVA (p-value = 0.159) and EVA (p-value = 0.303); CRI significantly dominates CEVA (p-value = 1.000) and EVA (p-value = 1.000). Remarkably, information differences versus cash flow always remain insignificant. To summarize, EBEI, RI, and CFO contain highest information content.

Results are consistent to previous evidence on the superiority of earnings, with earnings explaining 6% to 9% of stock returns.[478] Again, results do not confirm latest evidence on the outperformance of EVA by Feltham et al. (2004) and West and Worthington (2004). While Tsuji (2006) found high value-relevance for cash flows for 1982 to 2002, other studies commonly show low value-relevance of cash flow for the 1980s and 1990s.[479] Therefore, results may indicate that operating cash flows recently gained relevance to capital markets.

Moreover, negligible information content of EVA seems reasonable, due to the following: Using EVA data provided by Stern Stewart, previous research showed that EVA may explain firm value better than stock returns: E.g., O'Byrne showed that capitalized EVA explains 31% of the market value of the firm. Contrarily, other U.S. studies show that EVA explains only 2% to 5% of returns.[480] Additionally, analysis above yet showed that publicly available EVA data overestimate the value-relevance of EVA.

Consistent to above results, depreciation-adjusted gross metrics CRI and CEVA are outperformed by the net equivalents; and, CRI has comparable information content to leading measures. Consequently, results suggest that the relative information advantage of CVA over EVA refers to other components of the CVA metric than the depreciation adjustment.[481]

Table 29 summarizes results from four robustness tests, of which the first three were yet introduced above.

---

[478] See, e.g., Biddle et al. (1997), Chen and Dodd (2001), and Copland et al. (2004) finding that EBEI, operating income, and earnings per share explain 9%, 6%, and 6% of stock returns, respectively.

[479] See, e.g., Biddle et al. (1997), Feltham et al. (2004), O'Byrne (1996), and West and Worthington (2004).

[480] E.g., Biddle et al. (1997), Chen and Dodd (2001), Copeland et al. (2004), and Feltham et al. (2004) found that EVA explains 5%, 2%, 2%, and 4% of stock returns, respectively.

[481] See Günther et al. (1999) and Pälchen and Schremper (2001) for empirical studies comparing EVA and CVA.

Rank	(1)	>	(2)	>	(3)	>	(4)	>	(5)	>	(6)
A: Sign-based coefficients											
Measure	EBEI		RI		CFO		CRI		EVA		CEVA
Adj. R^2	0.122		0.104		0.096		0.094		0.018		0.016
B_1	0.564		1.012		0.695		1.386		0.383		0.394
B_2	1.351		1.250		0.048		0.942		0.239		0.182
B_3	-0.658		-0.355		-0.286		-0.483		-0.138		-0.173
B_4	0.617		0.552		-0.017		0.679		-0.105		-0.091
p-value		(0.000)		(0.000)		(0.000)		(1.000)		(1.000)	
				(0.000)		(1.000)		(0.000)		(1.000)	
						(0.644)		(0.724)		(0.000)	
								(0.000)		(0.868)	
										(0.003)	
B: Two-year return periods											
Measure	CFO		CRI		RI		EBEI		CEVA		EVA
Adj. R^2	0.107		0.023		0.022		0.018		0.008		0.007
p-value		(0.000)		(0.592)		(0.147)		(0.000)		(0.997)	
				(0.000)		(0.035)		(0.018)		(0.000)	
						(0.042)		(0.745)		(0.245)	
								(0.000)		(0.916)	
										(0.000)	
C: Five-year measurement periods											
Measure	RI		EBEI		CRI		CFO		EVA		CEVA
Adj. R^2	0.207		0.126		0.116		0.067		0.018		0.012
p-value		(0.000)		(0.971)		(0.000)		(0.237)		(1.000)	
				(0.998)		(0.178)		(0.622)		(0.406)	
						(0.000)		(1.000)		(0.287)	
								(0.787)		(1.000)	
										(0.564)	
D: Raw returns											
Measure	EBEI		RI		CFO		CRI		EVA		CEVA
Adj. R^2	0.100		0.086		0.077		0.074		0.005		0.005
p-value		(0.999)		(0.000)		(0.000)		(0.805)		(0.956)	
				(0.000)		(1.000)		(0.000)		(0.992)	
						(1.000)		(0.000)		(0.000)	
								(0.000)		(0.000)	
										(0.000)	

Table 29   Sensitivity Tests on Associations of Metrics with Stock Returns

First, constrained coefficients are replaced by sign-based coefficients.[482] As shown in Panel A of the table, adjusted R^2s substantially increase with sign-based coefficients, while the ranking and statements on significance largely remain: EBEI (adj. $R^2 = 12\%$) is leading, followed by RI and CFO (adj. $R^2 = 10\%$) and CRI, while all clearly outperform EVA and CEVA (adj. $R^2 = 2\%$). Consistent with prior studies, coefficients of positive performance levels and changes are larger than negative levels and changes.[483] All in all, this test further supports basic results.

The second modification includes a two-year return interval consisting of the contemporaneous and the subsequent year.[484] Allowing 15 rather than 3 months for information to be impounded in stock prices primarily aids the information contained by cash flows. As shown in Panel B, value-relevance of CFO increases considerably (adj. $R^2 = 10\%$), while value-relevance of EBEI, RI, and CRI decreases substantially (adj. $R^2 = 2\%$). Slight increases in adjusted R^2 for CEVA from 0.4% to about 0.8% do not provide sufficient support of the supposition that the market subsequently recognizes value-based performance measures, consistent with results from Biddle et al. (1997). Notably, CRI (p-value $= 0.592$) and CEVA (p-value $= 0.997$) gain significant information advantage over respective net equivalents. Consequently, this variation contests basic results.

Reducing errors from the expectation model, another regression model includes a five-year measurement interval.[485] Panel C shows a considerable increase in value-relevance specifically for earnings-based measures: RI (adj. $R^2 = 21\%$) is clearly leading, followed by EBEI (adj. $R^2 = 13\%$), CRI (adj. $R^2 = 12\%$), CFO (adj. $R^2 = 7\%$), EVA (adj. $R^2 = 2\%$), and, finally, CEVA (adj. $R^2 = 1\%$). Additionally, RI, EBEI, and CRI gain significant information advantage over other metrics. However, RI does not significantly outperform EBEI (p-value $= 0.000$). Consistent with findings of Biddle et al. (1997), these results do not confirm Stewart's claim that EVA shows the maximum information content on five-year data. Overall, results are in line with basic results.

---

[482] As yet shown in section 4.2.2, the underlying modified regression equation is

$$AR_t = \beta_0 + \beta_1 \cdot \frac{X_t^{pos}}{MVE_{t-1}} + \quad + \beta_2 \cdot \frac{(X_t - X_{t-1})^{pos}}{MVE_{t-1}} + \beta_3 \cdot \frac{X_t^{neg}}{MVE_{t-1}} + \beta_4 \cdot \frac{(X_t - X_{t-1})^{neg}}{MVE_{t-1}} + \varepsilon_t .$$

[483] Yet several studies provided evidence for earnings (Burgstahler & Dichev, 1997; Collins et al., 1999; Hayn, 1995) and residual income (Giner & Iniguez, 2006).

[484] As yet shown in section 4.2.2, the underlying modified regression equation is

$$(1 + AR_t) \cdot (1 + AR_{t+1}) - 1 =_t = \beta_0 + \beta_1 \cdot \frac{X_t}{MVE_{t-1}} + \beta_2 \cdot \frac{(X_t - X_{t-1})}{MVE_{t-1}} + \varepsilon_t .$$

[485] As yet shown in section 4.2.2, the underlying modified regression equation is

$$(1 + AR_t) \cdot (1 + AR_{t+1}) \cdot (1 + AR_{t+2}) \cdot (1 + AR_{t+3}) \cdot (1 + AR_{t+4}) - 1 = \beta_0 + \beta_1 \cdot \left(\frac{X_t}{MVE_{t-5}} + \frac{X_{t+1}}{MVE_{t-4}} + \frac{X_{t+2}}{MVE_{t-3}} + \right.$$

$$\left. + \frac{X_{t+3}}{MVE_{t-2}} + \frac{X_{t+4}}{MVE_{t-1}} \right) + \beta_2 \cdot \left(\frac{X_{t-4} - X_{t-5}}{MVE_{t-5}} + \frac{X_{t-3} - X_{t-4}}{MVE_{t-4}} + \frac{X_{t-2} - X_{t-3}}{MVE_{t-3}} + \frac{X_{t-1} - X_{t-2}}{MVE_{t-2}} + \frac{X_t - X_{t-1}}{MVE_{t-1}} \right) + \varepsilon_t .$$

Finally, another modification uses raw returns instead of abnormal return, to avoid flaws from incorrect abnormal return estimates.[486] As shown in the last panel of the following table, the ranking of measures largely remains, while information content increases: EBEI is leading (adj. R^2 = 10%), followed by RI (adj. R^2 = 9%), CFO (adj. R^2 = 8%), and CRI (adj. R^2 = 7%). Information content of EVA and CEVA (adj. R^2 = 0.5%) is negligible. Results are consistent with findings by Anastassis and Kyriazis (2007) and do not challenge basic results.

To summarize, the conducted sensitivity principally underlines fundamental results, since findings remained stable under three of four comparable modifications.

To this point, also basic regression results on associations with stock returns show remarkably low levels of adjusted R^2s, ranging from 1% to 8% (which are even lower than regression results on associations with firm value). Still under 5-year periods, adjusted R^2s reaching not more than 21% are still fairly low. However, additional analyses further support the soundness of results. E.g., earnings still dominate any residual income metric, when the regression model omits a proxy for unexpected performance.[487] Additionally, analysis showed that regression results from using the 'one-lag' model instead of the 'level and changes' model to approximate unexpected performance do not differ perceptibly.

5.2.3 Incremental Information Content: Importance of Earnings' Accruals

Again, the following regressions assume that levels of and changes in stock prices are determined by accounting measures. To examine incremental information content, changes in adjusted R^2s are evaluated.

Regression Analysis on Market Value

According to the second research hypothesis, unique information components of competing performance metrics (i.e., Accr, ATInt, CapChg, AccAdj, and DepAdj) are evaluated with respect to their contribution in explaining market values.[488] As shown in the previous chapter, one regression model is used to examine information components of CRI,[489] and another one is used to analyze information components of CEVA,[490] while both use CFO as base.

---

[486] The underlying modified regression equation is $RR_t = \beta_0 + \beta_1 \cdot \dfrac{X_t}{MVE_{t-1}} + \beta_2 \cdot \dfrac{(X_t - X_{t-1})}{MVE_{t-1}} + \varepsilon_t$, see section 4.2.2.

[487] E.g., an analysis of the selected 201 firms (after excluding extreme values above eight standard deviations and winsorizing data to 4 standard deviations) shows: CFO (adjusted R^2 = 4.9%) and EBEI (adjusted R^2 = 4.2%) dominate RI (adjusted R^2 = 1.0%), CRI (adjusted R^2 = 0.8%), EVA (adjusted R^2 = 0.2%), and CEVA (adjusted R^2 = 0.1%).

[488] See section 4.2.1 and 4.2.3.

[489] $\dfrac{MV_t}{BV_{t-1}} = \beta_0 + \beta_1 \cdot \dfrac{CFO_t / WACC_t}{BV_{t-1}} + \beta_2 \cdot \dfrac{Accr_t / WACC_t}{BV_{t-1}} + \beta_3 \cdot \dfrac{ATInt_t / WACC_t}{BV_{t-1}} + \beta_4 \cdot \dfrac{CapChg_t / WACC_t}{BV_{t-1}} + \beta_5 \cdot \dfrac{DepAdj_t^1 / WACC_t}{BV_{t-1}} + \varepsilon_t$

where CFO is the base, *Accr* are unique information components of EBEI, *ATInt* and *CapChg* are

Based on 2,128 firm-year observations, Table 30 shows changes in adjusted R^2 from sequential additions of the particular information component and respective p-values from two-tailed F-tests below in parentheses.

Sequence		(1)	(2)	(3)	(4-CRI)	(4-CEVA)	(5-CEVA)
Component	CFO	Accr	ATInt	CapChg	DepAdj1	AccAdj	DepAdj2
Δ Adj. R^2	0.097	0.119	0.012	0.000	0.003	0.008	0.003
p-value		(0.000)	(0.000)	(0.689)	(0.008)	(0.000)	(0.010)

Table 30   Incremental Information Content Tests referring to Firm Value

Differential information components are incrementally added to examine the information content of the 'last' distinctive information component, as follows: First, the incremental information content of earnings' accruals, *Accr*, equals the change in adjusted R^2 from regressing CFO and *Accr* with firm value versus regressing exclusively CFO with firm value.[491] Next, incremental information content of after-tax interest, *ATInt*, involves measuring the increase in adjusted R^2 from regressions results based on CFO, *Accr* and *ATInt*, on the one hand, and regression results exclusively based on CFO and *Accr*, on the other hand, etcetera. Results show that only earnings' accruals imply a considerable 12% increase in adjusted R^2 of 12%. Increases in adjusted R^2 due to addition of other components are marginal. However, all components except capital charges add significant incremental information content at the 1% level. Consequently, incremental results emphasize the value-relevance of earnings, while also pointing to the importance of the other metrics' differential information components.

Capital markets use primarily earnings to evaluate stock prices. In this respect, substantial (Δ adj. R^2 = 12%) and highly significant (p-value = 0.000) additional information conveyed by earnings' accruals seem to reflect the market's focus on earnings.

Since depreciation schedules are set several years in advance, accounting depreciation contains little 'surprise value'. Consequently, depreciation adjustments

---

unique information components of RI, and $DepAdj^1$ are unique information components of CRI (see section 4.2.2).

[490] Namely, $\dfrac{MV_t}{BV_{t-1}} = \beta_0 + \beta_1 \cdot \dfrac{CFO_t / WACC_t}{BV_{t-1}} + \beta_2 \cdot \dfrac{Accr_t / WACC_t}{BV_{t-1}} + \beta_3 \cdot \dfrac{ATInt_t / WACC_t}{BV_{t-1}} + \beta_4 \cdot \dfrac{CapChg_t / WACC_t}{BV_{t-1}} +$

$+ \beta_5 \cdot \dfrac{AccAdj_t / WACC_t}{BV_{t-1}} + \beta_5 \cdot \dfrac{DepAdj_t^2 / WACC_t}{BV_{t-1}} + \varepsilon_t$, where CFO is the base, *Accr* are unique information

components of EBEI, *ATInt* and *CapChg* are unique information components of RI, *AccAdj* are unique information components of EVA, and $DepAdj^2$ are unique information components of CEVA (see section 4.2.2).

[491] I.e., the two relevant regression functions are: $\dfrac{MV_t}{BV_{t-1}} = \beta_0 + \beta_1 \cdot \dfrac{CFO_t / WACC_t}{BV_{t-1}} + \beta_2 \cdot \dfrac{Accr_t / WACC_t}{BV_{t-1}} + \varepsilon_t$ and

$\dfrac{MV_t}{BV_{t-1}} = \beta_0 + \beta_1 \cdot \dfrac{CFO_t / WACC_t}{BV_{t-1}} + \varepsilon_t$.

that eliminate book depreciation are supposed to increase explanation of market value. Evidence on significant information conveyed by depreciation adjustments (p-value = 0.008 and 0.010) confirms suppositions on the information content of depreciation adjustments.

While Stewart argued that depreciation adjustments are insignificant, this analysis provides empirical evidence that refutes this claim.

Table 31 provides results from robustness tests, consistent to modifications applied above.[492]

Sequence		(1)	(2)	(3)	(4-CRI)	(4-CEVA)	(5-CEVA)
Component	CFO	Accr	ATInt	CapChg	DepAdj1	AccAdj	DepAdj2
A: Sign-based coefficients							
Δ Adj. R^2	0.113	0.111	0.011	0.000	0.003	0.018	0.002
p-value		(0.000)	(0.000)	(0.936)	(0.008)	(0.000)	(0.045)
B: Non-capitalized performance							
Δ Adj. R^2	0.160	0.108	0.025	0.004	0.000	0.008	0.000
p-value		(0.000)	(0.033)	(0.251)	(0.602)	(0.000)	(0.223)
C: Two-year market values							
Δ Adj. R^2	0.101	0.119	0.009	0.004	0.002	0.008	0.001
p-value		(0.000)	(0.002)	(0.037)	(0.036)	(0.000)	(0.059)
D: Five-year changes							
Δ Adj. R^2	0.027	0.077	0.001	0.040	0.022	0.003	0.020
p-value		(0.000)	(0.548)	(0.000)	(0.000)	(0.099)	(0.000)

Table 31   Sensitivity Tests on Associations of Components with Firm Value

First, incremental information content tests are repeated with sign-based coefficients. As shown in Panel A of the Table 31, results largely prevail: Only accruals substantially (Δ adj. R^2 = 11%) and significantly (p-value = 0.000) aid the explanation of firm values, while other components provide only marginal incremental information content. Results on significance are unchanged, except that the significance level of CEVA's depreciation adjustments decreases to 5%.

Second, analysis is repeated with non-capitalized components, to avoid perpetuity assumptions. Results, see Panel B, reveal that only accruals and accounting adjustments maintain significant incremental information content at the 1% level (p-value = 0.000). Again, accruals is the only component with substantial incremental information content (Δ adj. R^2 = 11%).

Third, components are related to the firm value of the contemporaneous and subsequent fiscal year. As shown in Panel C, all components have significant incremental information content at a 10% significance level. Again, only accruals provide sizeable incremental information content (Δ adj. R^2 = 12%).

---

[492] See section 5.2.2.

Based on 5-year changes, see Panel D, all components except after-tax interest expense have significant incremental information content at the 10% level. Further, findings demonstrate an increase in the additional information conveyed by depreciation adjustments (Δ adj. $R^2 = 2\%$).

To conclude, various model extensions confirm that only the accruals component contains considerable incremental information content.

Regression Analysis on Stock Returns

According to the fourth research hypothesis, unique information components of competing performance metrics (Accr, ATInt, CapChg, AccAdj, and DepAdj) are evaluated as to what they contribute to the explanation of stock returns. Again, using CFO as a base, two regression models are used to assess information components of CRI and CEVA.[493]

Based on 1,922 firm-year observations, Table 32 summarizes respective results.

Sequence		(1)	(2)	(3)	(4-CRI)	(4-CEVA)	(5-CEVA)
Component	CFO	Accr	ATInt	CapChg	DepAdj1	AccAdj	DepAdj2
Δ Adj. R^2	0.074	0.046	0.016	0.006	0.011	0.006	0.009
p-value		(0.000)	(0.032)	(0.174)	(0.005)	(0.240)	(0.016)

Table 32   Incremental Information Content Tests referring to Stock Returns

Only the accruals component of EBEI leads to a substantial (Δ adj. $R^2 = 5\%$) and highly significant (p-value = 0.000) improvement in explaining stock returns, in line with findings from Biddle et al. (1997). Other components imply only marginal additions to value-relevance. Nevertheless, after-tax interest and depreciation adjustments still contain significant incremental information content at the 5% level.

---

[493] As yet introduced in section 4.2.2, the following equation allows assessing components of CRI:

$$AR_t = \beta_0 + \beta_1 \cdot \frac{CFO_t}{MVE_{t-1}} + \beta_2 \cdot \frac{(CFO_t - CFO_{t-1})}{MVE_{t-1}} + \beta_3 \cdot \frac{Accr_t}{MVE_{t-1}} + \beta_4 \cdot \frac{(Accr_t - Accr_{t-1})}{MVE_{t-1}} + \beta_5 \cdot \frac{ATInt_t}{MVE_{t-1}} + \beta_6 \cdot$$

$$\frac{(ATInt_t - ATInt_{t-1})}{MVE_{t-1}} + \beta_7 \cdot \frac{CapChg_t}{MVE_{t-1}} + \beta_8 \cdot \frac{(CapChg_t - CapChg_{t-1})}{MVE_{t-1}} + \beta_9 \cdot \frac{DepAdj_t^1}{MVE_{t-1}} + \beta_{10} \cdot \frac{(DepAdj_t^1 - DepAdj_{t-1}^1)}{MVE_{t-1}} + \varepsilon_t$$

but then, the following alternate equation is used for evaluating components of CEVA: $AR_t = \beta_0 +$

$$+ \beta_1 \cdot \frac{CFO_t}{MVE_{t-1}} + \beta_2 \cdot \frac{(CFO_t - CFO_{t-1})}{MVE_{t-1}} + \beta_3 \cdot \frac{Accr_t}{MVE_{t-1}} + \beta_4 \cdot \frac{(Accr_t - Accr_{t-1})}{MVE_{t-1}} + \beta_5 \cdot \frac{ATInt_t}{MVE_{t-1}} +$$

$$+ \beta_6 \cdot \frac{(ATInt_t - ATInt_{t-1})}{MVE_{t-1}} + \beta_7 \cdot \frac{CapChg_t}{MVE_{t-1}} + \beta_8 \cdot \frac{(CapChg_t - CapChg_{t-1})}{MVE_{t-1}} + \beta_9 \cdot \frac{AccAdj_t}{MVE_{t-1}} +$$

$$+ \beta_{10} \cdot \frac{(AccAdj_t - AccAdj_{t-1})}{MVE_{t-1}} + \beta_{11} \cdot \frac{DepAdj_t^2}{MVE_{t-1}} + \beta_{12} \cdot \frac{(DepAdj_t^2 - DepAdj_{t-1}^2)}{MVE_{t-1}} + \varepsilon_t$$, where CFO is the

base, *Accr* are unique information components of EBEI, *ATInt* and *CapChg* are unique information components of RI, *DepAdj¹* are unique information components of CRI, *AccAdj* are unique information components of EVA, and *DepAdj²* are unique information components of CEVA.

Table 33 provides sensitivity tests of the basic specification tested above, based on model variations introduced above.[494]

Sequence		(1)	(2)	(3)	(4-CRI)	(4-CEVA)	(5-CEVA)
Component	CFO	Accr	ATInt	CapChg	DepAdj1	AccAdj	DepAdj2
A: Sign-based coefficients							
Δ Adj. R^2	0.096	0.047	0.002	0.001	0.010	0.007	0.008
p-value		(0.001)	(0.611)	(0.684)	(0.042)	(0.682)	(0.141)
B: Two-year return periods							
Δ Adj. R^2	0.107	0.003	0.025	0.012	0.003	-0.001	0.002
p-value		(0.312)	(0.001)	(0.006)	(0.185)	(0.648)	(0.196)
C: Five-year measurement periods							
Δ Adj. R^2	0.096	0.110	0.022	0.017	0.006	0.017	-0.002
p-value		(0.001)	(0.000)	(0.241)	(0.126)	(0.038)	(0.811)
D: Raw returns							
Δ Adj. R^2	0.080	0.060	0.014	0.012	0.010	0.007	0.010
p-value		(0.000)	(0.070)	(0.038)	(0.010)	(0.266)	(0.009)

Table 33   Sensitivity Tests on Associations of Components with Stock Returns

Results largely prevail under sign-based coefficients, see Panel A: Accruals are the only component with substantial incremental information content (Δ adj. R^2 = 5%), while other components than depreciation adjustments of CRI convey marginal and insignificant additional information. Depreciation adjustments included in the CRI measure provide marginal (Δ adj. R^2 = 1%) but significant information content (p-value = 0.042).

Results change substantially based on 2-year average returns, see Panel B: No component implies a substantial increase in value-relevance. Moreover, only RI components contain significant incremental information content, at a 1% significance level. Consequently, it seems that unexpected performance is quickly impounded into stock prices.

Analysis of 5-year periods, see Panel C, further supports the clear outperformance of accruals: Accruals contain substantial (Δ adj. R^2 = 11%) and highly significant (p-value = 0.001) incremental information content. Accounting adjustments are gaining significant information content at the 5% level (p-value = 0.038), while depreciation adjustments have no longer significant information content. Therefore, it seems that a longer time period overcomes potential errors from the expectancy model.

Finally, associations with raw returns, see Panel D, additionally support basic results: Again, accruals contain substantial (Δ adj. R^2 = 6%) and highly significant (p-value = 0.000) incremental information content. Next to depreciation adjustments and

---

[494] See section 4.2.2.

after-tax interest, capital charges additionally provide significant incremental information (p-value = 0.038).

Therefore, sensitivity tests again reconfirm basic results.

Altogether, results from relating components to stock returns are largely consistent with results from relating components to market value. Furthermore, the higher incremental information content of accruals with respect to the explanation of market value (Δ adj. R^2 = 12%) than with respect to the explanation of stock returns (Δ adj. R^2 = 5%) is in line with relative results, described in the previous section.

5.3 Summary

The variables of the following empirical study (that are EBEI, CFO, RI, EVA, CRI, and CEVA as independent variables and MV and AR as dependent variables) have been first calculated and reviewed for a well-known international company (the Campbell Soup Corporation). Then, final results were discussed for Campbell Soup Corporation (as base case), Alaska Air Group (representing a firm with relatively young assets), and Unifirst (representing a firm with relatively old assets). The case study shows the following patterns for the variables:

CFO is particularly high and fairly volatile. Average EBEI is much lower than average CFO, due to non-cash expenses, and considerably less volatile, due to smoothing accruals and reserves. Average RI is much lower than average EBEI, since it additionally includes charges for equity capital. EVA differs substantially from RI and is substantially more volatile, driven by a wide set of applied accounting adjustments. For Unifirst that carries relatively old assets, CRI and CEVA (and likewise the profitability measures ROGA and CROCE) are consistently lower than RI and EVA, respectively, reducing underestimations of performance due to book depreciation; for Alaska Air that maintains a relatively young asset basis, CRI and CEVA (and also ROGA and CROCE) are constantly higher than RI and EVA, respectively, reducing overestimates of value creation due to accounting depreciation. Correlations of accounting metrics with market measures vary considerably.

Moreover, results from the case study may at first glance wrongly suggest that depreciation adjustments (avoiding built-in performance improvements over the asset life), as included in CRI and CEVA, are worthless, since firms commonly have a relatively stable average age of their assets over time, replacing assets regularly. However, it has to be noted that also in this case VBM incentive systems based on RI or EVA would favor distorted investment management decisions, providing incentives to squeeze investments beyond strategically optimal levels. On the contrary, CRI and CEVA avoid underestimation (and overestimation) of measured performance in the first (and last) years of the asset life.

Next, the empirical section of this study examined the value-relevance of fundamental value-based measures, RI and EVA, relative to depreciation-adjusted equivalents, CRI and CEVA, and currently mandated earnings and cash flow measures, EBEI and CFO. On the basis of a sample consisting of 2,147 observations from U.S. firms, this study provides the following relative and incremental results:

Analysis of competing measures supports a slight advantage of earnings and RI, that are highly related to firm value (adj. R^2 = 19%) and stock returns (adj. R^2 = 8%), although such regression results explain only part of the overall stock price that remains dominated by other 'soft' factors driving investors' expectations. However, detection of other factors than accounting performance that are useful to explain stock prices is beyond the scope of this study. CRI still explains 17% and 6% of market value and returns, respectively, and is not significantly outperformed by earnings. However, value-relevance of EVA and CEVA is comparatively low, especially when related to stock returns. Moreover, RI and EVA significantly outperform respective depreciation-adjusted equivalents CRI and CEVA. Finally, operating cash flow is less value-relevant than EBEI and RI.

Incrementally, evaluation of unique information components (characterizing the respective accounting-based performance metric) reveals that the additional information of earnings' accruals towards the explanation of market value (Δ adj. R^2 = 12%) and stock returns (Δ adj. R^2 = 5%) is substantial and highly significant (p-value = 0,000). Although depreciation adjustments of CRI and CEVA just cause increases in adjusted R^2 of 0.3% and 1% relating to firm value and return, respectively, marginal incremental information is significant at the 5% level. Consequently, respective additions of the depreciation adjustment component to value-relevance is significant but yet too marginal to cause dominance of CRI or CEVA over net earnings. Likewise, accounting adjustments of EVA are just marginally (Δ adj. R^2 = 1%) adding to the explanation of firm value or stock returns, being significant only when related to market value.

To conclude, empirical results are in line with prior evidence on the dominance of earnings. This means latest findings on the superiority of EVA are not reconfirmed. While also newly introduced measures CRI and CEVA do not outperform earnings, CRI contains high relative information content and both CRI and CEVA contain marginal but significant incremental information content. Therefore, claims of Stewart on the insignificance of depreciation adjustments are refuted. Contrary to previous studies that use EVA data considering only few accounting adjustments (mostly publicly available EVA data provided by Stern Stewart), this study finds considerably lower value-relevance of EVA, using independently calculated EVA data that includes a wide set of adjustments. Consequently, results indicate that value-relevance of EVA has been so far overestimated.

6 Discussion

Motivated by age bias inherent in net residual income measures RI and EVA and addressing potential interest among executive managers, investors and accounting policy makers, this research developed two new value-based performance measures CRI and CEVA and tested the value-relevance of common and newly introduced performance measures. While the new measures and empirical results were yet presented in the previous chapters, this chapter deals with the implications and concludes with limitations of this study and suggestions for future research.

6.1 Implications

Some conclusions can be drawn from both the methodological and statistical results of this study.

From a methodological view, this research explicates the advantages of the two newly developed CRI and CEVA value metrics. Contrary to disclosed and commonly used book earnings, CRI and CEVA constitute depreciation-adjusted VBM metrics, which minimize bias induced by accounting depreciation and are conceptually superior measures of the firm's value creation (conforming to valuation theory).

In fact, age bias due to accounting depreciation plays a major role from two views: Under increases (or decreases) in the age of the asset, there are built-in performance improvements (or deteriorations) over time, independent of economic value creation. Equally important, in the first (and last) years of the asset life, measured performance is underestimated (and overestimated), with respect to the 'true' level of shareholder value added. In this respect, a depreciation-adjusted measure such as CRI and CEVA provides clear advantages, if applied, e.g., to firms, business units, and/or geographic divisions with significant changes in average residual asset lives and/or with relatively new or old equipment.

On this basis, firms should disclose CRI and/or CEVA results in their financial statements, to provide value-based information on the firm's performance, undistorted by book depreciation. This should be in particular referred to firms presenting a high growth momentum of capital employed and/or acting in sectors where high investment replacement rates are required, as in these cases RI and EVA may significantly underestimate the value performance of new investments and this may affect management's and investors' strategic decisions. Disclosing supplemental information on CRI and/or CEVA and underlying key calculation components within the consolidated income statements, the operations review, or the notes to the consolidated financial statements would enable:

- to improve the quality and relevance of financial statements, from a value-based perspective,

- to mitigate management's investment bias under earnings, RI, and EVA based on an externally verified metric, given audited CRI or CEVA is adopted for management compensation and performance measurement,

- to communicate the economic value creation of the firm to employees and, above all, investors, directors, analysts, and other stakeholders and interest groups, avoiding distortions from accounting depreciation,

- to implement a transparent, accountable and improved value-based management incentive system, and

- to externally monitor the firm performance or select portfolios, using a depreciation-neutral value-based approach.

From a statistical perspective, empirical results reconfirm the earnings myopia of financial markets and the possible predominance of other non-accounting 'soft' aspects driving investors' expectations. Moreover, low value-relevance of independently and exhaustively estimated EVA values provides some evidence for a prior overestimation of EVA's value-relevance. Finally, regression results identify CRI as practical value-based and depreciation-adjusted performance metric that is leading in explaining shareholder value.

6.2 Limitations

The interpretation of the empirical results requires to point towards several constraints. In particular, to understand low value-relevance for CRI and CEVA, it is important to recognize four limitations:

- Markets may have not recognized CRI and CEVA, since they have not been published through the underlying period. Consistent with the concept of 'earnings myopia', the maintained hypothesis of semi-strong market efficiency may only apply to disclosed earnings or closely related measures.

- Accounting depreciation may provide useful information to investors, signaling forthcoming cash management or utility issues, or concerns regarding future shareholder risks and returns, reflected by earnings' superior value-relevance. Nevertheless, a depreciation-adjusted measure, such as CRI and CEVA, may indeed provide a better measure of 'true' economic value created. Equivalently, undisclosed RI and EVA may have low value-relevance due to the earnings myopia of markets but, nevertheless, provide useful economic information regardless of its value-relevance.

- Age bias from book depreciation may not considerably affect performance schedules of large firms, carrying a portfolio of assets that are constantly replaced, thus, maintaining a constant average asset life and avoiding a prevalence of new or old assets.

- Underlying estimates of the depreciation adjustment may be different from those used by the market. Specifically, data may imply measurement errors of asset lives due to a lack of disclosures from SFAS 39, par. 5, requiring solely disclosure of major asset class balances and respective ranges of asset lives.

Above all, value-relevance of residual income measures, i.e., RI, EVA, CRI, and CEVA, is affected by the following main restraints:

- There may be errors in the estimates of the capital charge relative to the market, since the applied Capital Asset Pricing Model and included beta estimates look 'backwards', using historical parameters to calculate the cost of equity.

- Additionally, accounting adjustments included in EVA and CEVA may involve measurement errors relative to the market.

- The assumption of the cross-sectional returns paradigm to have constant regression coefficients for all firms and sectors may not hold.

6.3 Outlook

This study has newly introduced the CRI and CEVA metric. However, another more sophisticated new metric can be developed from these newly introduced metrics and from the concepts included in this study: this new metric, that could be called 'Market Economic Value Added' (MEVA), can provide the value creation with respect to any asset value other than gross or net assets. It can be used for analyzing the value creation in respect of the market value of the firm at a certain date or, when it is convenient, to refer to a certain asset value that, for example, incorporates past value decreases (write-downs) or expected value creation components already included in the firm's stock market quotation at a certain date.

Furthermore, there are at least three ways to extend this research. First, a study may examine the performance of firms that have implemented an unbiased value-based measure such as CRI and CEVA. It is conceivable that associations are stronger for adopters that publish their measures, since the market may be sufficiently efficient to recognize 'true' value creation if sufficiently disclosed. A further avenue of research is the change of the unit of analysis to a single project or organizational unit; for instance, research may address the central issues of optimal compensation design using CEVA or CRI, depending on the project type or hierarchical and divisional level. Finally, future research shall consider other usefulness criteria than the prediction of the stock price for evaluating a depreciation-adjusted value-based performance measures, such as qualitative characteristics of managerial activity with respect to the reduction of agency costs.

References

Abate, J. A., & Grant, J. L. (2001). *Focus on Value: A Corporate and Investor Guide to Wealth Creation*. New York: John Wiley & Sons.

Abate, J. A., Grant, J. L., & Stewart, G. B. (2004). The EVA Style of Investing. *The Journal of Portfolio Management, Summer*, 61-72.

Aders, C. (2001). *Value Based Management: Vom Discounted Cashflow zu den operativen Werttreibern*. Paper presented at the Munich Workshop for Accounting and Auditing at the LMU München, 26. November 2001.

Ali, A., & Pope, P. F. (1995). The incremental information content of earnings, funds flow and cash flow: The UK evidence. *Journal of Business Finance & Accounting, 22*(1), 19-34.

Anastassis, C., & Kyriazis, D. (2007). The Validity of the Economic Value Added Approach: an Empirical Application. *European Financial Management, 13*(1), 71-100.

Anderson, A. M., Bey, R. P., & Weaver, S. C. (2004). *Economic Value Added® Adjustments: Much to Do About Nothing?* Paper presented at the Financial Management Association Annual Conference.

Anderson, C. (1997). Values-based management. *Academy of Management Executive, 11*, 25-46.

Antill, N., & Arnott, R. (2004). Creating Value in the Oil Industry. *Journal of Applied Corporate Finance, 16*(1), 18-31.

Ashworth, G. (2000). *Delivering Shareholder Value through Integrated Performance Management* (1st ed.): FT Prentice Hall.

Bacidore, J. M., Boquist, J. A., Milbourn, T. T., & Thakor, A. V. (1997). The search for the best financial performance measure. *Financial Analysts Journal, 53*(3), 11-20.

Ball, R., & Brown, P. (1968). An Empirical Evaluation of Accounting Income Numbers. *Journal of Accounting Research, 6*(2), 159-178.

Barber, B. M., & Lyon, J. D. (1996). Detecting Abnormal Operating Performance: The Empirical Power and Specification of Test-Statistics. *Journal of Financial Economics, 15*, 61-89.

Beaver, W. H. (1968). The Information Content of Annual Earnings Announcements. *Journal of Accounting Research*(Supplement), 67-92.

Berle, A. A., & Means, G. C. ([1932] 1968). *The Modern Corporation and Private Property*. New York: Harcourt, Brace & World.

Bernard, V. L. (1994). The Feltham-Ohlson Framework: Implications for Empiricists. *Contemporary Accounting Research, Spring*, 733-747.

Bernard, V. L., & Stober, T. L. (1989). The Nature and Amount of Information in Cash Flows and Accruals. *The Accounting Review, 64*(4), 624-652.

Bernstein, R., & Subramanian, S. (2003, 27 January 2003). An Analysis of CFROI®. *Quantitative Viewpoint*.

Biddle, G. C., Bowen, R. M., & Wallace, J. S. (1997). Does EVA beat earnings? Evidence on associations with stock returns and firm values. *Journal of Accounting & Economics, 24*(3), 301-336.

Biddle, G. C., Bowen, R. M., & Wallace, J. S. (1999). Evidence on EVA. *Journal of Applied Corporate Finance, 12*(2), 69-79.

Biddle, G. C., Chen, P., & Zhang, G. (2001). When Capital Follows Profitability: Non-linear Residual Income Dynamics. *Review of Accounting Studies, 6*, 229-265.

Biddle, G. C., Seow, G., & Siegel, A. (1995). Relative versus incremental information content. *Contemporary Accounting Research, 12*, 1-23.

Biddle, G. C., & Seow, G. S. (1991). The Estimation and Determinants of Associations Between Returns and Earnings: Evidence from Cross-industry Comparisons. *Journal of Accounting, Auditing & Finance, 6*(2), 183-232.

Black, E., Carnes, T., & Richardson, V. (2000). The Market Valuation of Firm Reputation. *The Corporate Reputation Review, 3*(1), 31-42.

Boston Consulting Group. (1994). *Shareholder Value Management: Improved Measurement Drives Improved Value Creation* (Vol. 2). Boston: Boston Consulting Group.

Boston Consulting Group. (1999). *The Value Creators: A Study of The World's Top Performers*. Boston: Boston Consulting Group.

Boston Consulting Group. (2000a). *New perspectives on value creation: a study of the world's top performers*. Boston: Boston Consulting Groupo. Document Number)

Boston Consulting Group. (2000b). *The Value Creators: New Perspectives on Value Creation*. Boston: Boston Consulting Group.

Boston Consulting Group. (2001). *The Value Creators: Dealing with Investor's Expectations*. Boston: Boston Consulting Group.

Boston Consulting Group. (2002). *The Value Creators: Succeed in Uncertain Times*. Boston: Boston Consulting Group.

Boston Consulting Group. (2003). *The Value Creators: Back to Fundamentals*. Boston: Boston Consulting Group.

Boston Consulting Group. (2004). *The Value Creators: Next Frontier*. Boston: Boston Consulting Group.

Boston Consulting Group. (2005). *The Value Creators: Balancing Act*. Boston: Boston Consulting Group.

Boston Consulting Group. (2006). *The Value Creators: Spotlight on Growth*. Boston: Boston Consulting Group.

Boston Consulting Group. (2007). *The Value Creators: Avoiding the Cash Trap*. Boston: Boston Consulting Group.

Boston Consulting Group. (2008). *The Value Creators: The Missing Link.* Boston: Boston Consulting Group.

Bowen, H. R. (1953). *Social Responsibilities of the Businessman.* New York: Harper & Brothers.

Bowen, R. M., Burgstahler, D., & Daley, L. A. (1987). The Incremental Information Content of Accrual Versus Cash Flows. *The Accounting Review, 62*(4), 723-747.

Bowen, R. M., Johnson, M. F., & Shevlin, T. (1989). Informational Efficiency and the Information Content of Earnings During the Market Crash of October 1987. *Journal of Accounting & Economics, 11*(2/3), 225-254.

Bowen, R. M., & Wallace, J. S. (1999). Interior Systems. *Issues in Accounting Education, 14* (3), 517-541.

Bradley, M., & Jarrell, G. A. (2008). Expected Inflation and the Constant-Growth Valuation Model. *Journal of Applied Corporate Finance, 20*(2), 66-78.

Brealey, R. A., & Myers, S. C. (2000). *Principles of Corporate Finance* (6th international ed.). Boston: Irwin McGraw-Hill.

Bruntland, G. H. (1987). *Our Common Future.* Oxford: Oxford University Press.

Burgstahler, D. C., & Dichev, I. D. (1997). Earnings, Adaptation and Equity Value. *The Accounting Review, 72*(2, April), 187-215.

Calandro, J., Jr., & Lane, S. (2002). The Insurance Performance Measure: Bringing Value to the Insurance Industry. *Journal of Applied Corporate Finance, 14*(4), 94-99.

Canning, J. B. (1929). *The Economics of Accountancy.* New York: Ronald Press.

Chen, S., & Clinton, B. D. (1998). Do Performance Measures Measure Up? *Management Accounting: Official Magazine of Institute of Management Accountants, 80*(4), 38-42.

Chen, S., & Dodd, J. L. (1997). Economic Value Added (EVA™): An Empricial Examination Of A New Corporate Performance Measure. *Journal of Managerial Issues, 9*(3), 318-333.

Chen, S., & Dodd, J. L. (2001). Operating Income, Residual Income and EVA™: Which Metric Is More Value Relevant? *Journal of Managerial Issues, 13*(1), 65-86.

Cheng, C. S. A., Cheung, J. K., & Gopalakrishnan, V. (1993). On the Usefulness of Operating Income, Net Income and Comprehensive Income in Explaining Security Returns. *Accounting and Business Research, 23*(91), 195-203.

Chew, D. C., Stern, J. M., & Stewart, G. B. (1995). The EVA Financial System. *Journal of Applied Corporate Finance, 8*(2), 32-46.

Chew, D. C., Stern, J. M., & Stewart, G. B. (1996). EVA®: An integrated financial management system. *European Financial Management, 2*(2), 223-245.

Christie, A. A. (1987). On cross-sectional analysis in accounting research. *Journal of Accounting & Economics, 9*(3), 231-258.

Collins, D. W., Pincus, M., & Xie, H. (1999). Equity Valuation and Negative Earnings: The Role of Book Value of Equity. *The Accounting Review, 74*(1), 29-61.

Colvin, G. (2000, December 18, 2000). America's Best & Worst Wealth Creators. *Fortune*, 207-216.

Copeland, T., Dolgoff, A., & Moel, A. (2004). The Role of Expectations in Explaining the Cross-Section of Stock Returns. *Review of Accounting Studies, 9*, 149-188.

Copeland, T., Koller, T., & Murrin, J. (1994). *Valuation: Measuring and Managing the Value of Companies* (2nd ed.). New York: John Wiley & Sons.

Copeland, T., Koller, T., & Murrin, J. (2000). *Valuation: Measuring and Managing the Value of Companies* (University ed.). New York: John Wiley & Sons.

Cordeiro, J. J., & Kent, D. D. J. (2001). Do EVA^{TM} Adopters Outperform their Industry Peers? Evidence from Security Analyst Earnings Forecasts. *American Business Review, 19*(2), 57-63.

Cox, D. (1961). *Tests of separate families of hypotheses.* Paper presented at the Fourth Berkeley Symposium on Mathematical Statistics and Probability, Berkeley.

Crasselt, N. (2001). Rappaports Shareholder Value Added. *Finanz Betrieb, 3*, 165-171.

Davis , E., Flanders, S., & Star, J. (1991). Who Are the World's Most Successful Companies? *Business Strategy Review, 2*(2), 1-33.

Davis, E., & Kay, J. (1990). Assessing Corporate Performance. *Business Strategy Review, 1*(2), 1-16.

Day, G. S., & Fahey, L. (1990). Putting Strategy into Shareholder Value Analysis. *Harvard Business Review, 68*(2), 156-162.

Dimson, E. (1979). Risk Measurement When Shares Are Subject to Infrequent Trading. *Journal of Financial Economics, 7*(2), 197-226.

Dodd, J. L., & Johns, J. (1999). EVA Reconsidered. *Business & Economic Review, 45*(3), 13-18.

Donaldson, T., & Preston, L. E. (1995). The Stakeholder Theory of the Corporation: Concepts, Evidence, and Implications. *Academy of Management Review, 20*(1), 65-91.

Drucker, P. F. (1995). The information executives truly need. *Harvard Business Review, 73*(1), 54-62.

Easton, P. D., & Harris, T. S. (1991). Earnings as a Explanatory Variable for Returns. *Journal of Accounting Research, 29*(1), 19-36.

Edey, H. C. (1957). Business valuation, goodwill and the super-profit method. *Accountancy, January/February*.

Edwards, E. O., & Bell, P. W. (1961). *The Theory and Measurement of Business Income*: University of California Press.

Ehrbar, A. (1998). *EVA: the real key to creating wealth*. New York: Wiley.

Elkington, J. (1997). *Cannibals with Forks: The Triple Bottom Line of 21st Century Business*. Oxford: Capstone Publishing.

Erasmus, P. (2008). Value Based Financial Performance Measures: An Evaluation of Relative and Incremental Information Content. *Corporate Ownership & Control, 6*(1), 66-77.

Fama, E. F. (1998). Market Efficiency, Long-Term Returns, and Behavioral Finance. *Journal of Financial Economics, 49*(3), 283-306.

Fama, E. F., & French, K. R. (1992). The Cross-Section of Expected Returns. *Journal of Finance, 47*(2), 427-465.

Feltham, G. A., & Ohlson, J. A. (1995). Valuation and Clean Surplus Accounting for Operating and Financial Activities. *Contemporary Accounting Research, 11*(2), 689-731.

Feltham, G. D., Isaac, G. E., Mbagwu, C., & Vaidyanathan, G. (2004). Perhaps EVA Does Beat Earnings - Revisiting Previous Evidence. *Journal of Applied Corporate Finance, 16*(1), 83-88.

Fernández, P. (2001). EVA and Cash Value Added Do Not Measure Shareholder Value Creation [Electronic Version]. *Social Science Research Network Electronic Paper Collection, June*, from http://www.ssrn.com/

Fernández, P. (2002a). *EVA, Economic profit and cash value added do NOT measure shareholder value creation.*Unpublished manuscript, IESE Business School Research Paper D/453.

Fernández, P. (2002b). *Valuation Methods and Shareholder Value Creation.* Burlington, MA: Academic Press.

Fidell, L. S., & Tabachnick, B. G. (2001). *Using multivariate statistics* (4th ed.). Boston: Allyn and Bacon.

Finegan, P. T. (1991). Maximizing Shareholder Value at the Private Company. *Journal of Applied Corporate Finance, 4*, 30-45.

Fischer, T. M., & Rödl, K. (2003). Strategische und wertorientierte Managementkonzepte in der Unternehmenspublizität - Analyse der DAX 30-Geschäftsberichte in einer unternehmenskulturellen Perspektive. Lehrstuhl für ABWL, Controlling und Wirtschaftsprüfung, Katholische Universität Eichstätt-Ingolstadt.

Fletcher, H. D., & Smith, D. B. (2004). Managing For Value: Developing a Performance Measurement System Integrating Economic Value Added and the Balanced Scorecard in Strategic Planning. *Journal of Business Strategies, 21*(1), 1-17.

Foster, G. (1986). *Financial Statement Analysis*. Englewood Cliffs, NJ: Prentice-Hall.

Frankel, R., & Lee, C. M. C. (1998). Accounting Valuation, Market Expectation, and Cross-sectional stock returns. *Journal of Accounting and Economics, 25*(3), 283-319.

Freeman, R. E. (1984). *Strategic Management: A stakeholder approach*. Boston: Pitman.

Frigo, M. L. (2002). Strategy Execution and Value-Based Management. *Strategic Management, October*, 6-9.

Gandhok, T., Dwivedi, A., & Lal, J. (2001). EVAluating Mergers and Acquisitions - How to avoid overpaying. *EVAluation (by Stern Stewart & Co.), 3*, 1-8.

Giner, B., & Iniguez, R. (2006). An empirical assessment of the Feltham-Ohlson models considering the sign of abnormal earnings. *Accounting and Business Research, 36*(3), 169-190.

Grant, J. L. (2003). *Foundations of economic valued added*. New Jersey: John Wiley & Sons.

Griffith, J. M. (2004). The True Value of EVA^R. *Journal of Applied Finance, 14*(2), 25-29.

Günther, T., Landrock, B., & Muche, T. (1999). *Profit versus Value-Based Performance Measures – An Empirical Investigation Based on the Correlation with Capital Market Returns for German DAX-100 Companies.*Unpublished manuscript, Dresdner Beiträge zur Betriebswirtschaftslehre 23/99, 1-31.

Guthrie, G. L., & Lemon, L. D. (2004). *Mathematics of Interest Rates and Finance.* Upper Saddle River, NJ: Prentice Hall.

Hamilton, R. (1777). *An Introduction to Merchandize.* Edinburgh.

Haspeslagh, P., Noda, T., & Boulos, F. (2001). It's Not Just About the Numbers. *Harvard Business Review, July-August,* 64-73.

Hayn, C. (1995). The information content of losses. *Journal of Accounting & Economics, 20,* 125-153.

Heidorn, T., Siebrecht, F., & Klein, H.-D. (2001). Economic Value Added zur Erklärung der Bewertung europäischer Aktien. *Finanzbetrieb, Oktober,* 560-564.

Henderson, D. (2001). *Misguided Virtue. False Notions of Corporate Social Responsability.* Paper presented at the New Zealand Business RoundTable, Wellington.

Herzberg, M. M. (1998). Implementing EBO/EVA Analysis in Stock Selection. *The Journal of Investing, Spring,* 45-53.

Hodak, M. (2000). The Viable EVA Center (Or, How to Slice a Company So It Doesn't Bleed). *Journal of Applied Corporate Finance, 13*(3), 71-79.

Hogan, C. E., & Lewis, C. M. (2005). Long-Run Investment Decisions, Operating Performance, and Shareholder Value Creation of Firms Adopting Compensation Plans Based on Economic Profits. *Journal of Financial and Quantitative Analysis, 40*(4), 721-745.

Holthausen, R. W., & Watts, R. L. (2001). The relevance of the value-relevance literature for financial accounting standard setting. *Journal of Accounting & Economics, 31,* 3-75.

Hostettler, S. (1997). *Das Konzept des Economic Value Added (EVA) - Massstab für finanzielle Performance und Bewertungsinstrument im Zeichen des Shareholder Value - Darstellung und Anwendung auf Schweizer Aktiengesellschaften.* Bern: Paul Haupt.

How the Scoreboard Ranking Were Compiled. (2008). *Wall Street Journal - Eastern Edition, 251*(45), R7.

Hunter, D. M., & Simon, D. P. (2005). A Conditional Assessment of the Relationships between the Major World Bond Markets. *European Financial Management, 11*(4), 463–482.

Ittner, C. D., & Larcker, D. F. (2001). Assessing empirical research in managerial accounting: a value-based management perspective. *Journal of Accounting & Economics, 32*(1-3), 349-410.

Jackson, A. (1996). The How and Why of EVA AT CS First Boston. *Journal of Applied Corporate Finance, 9*(1), 98-103.

Jacobson, R. (1987). The Validity of ROI as a Measure of Business Performance. *The American Economic Review, 77*(3), 470-478.

Jacquillat, B., & Solnick, B. (1978). Multinationals are poor tools for diversification. *Journal of Portfolio Management, 4*(2), 8-12.

Jensen, M. C., & Meckling, W. H. (1990). Specific Knowledge and Divisional Performance Measurement. *Journal of Applied Corporate Finance, 12*(2), 8-17.

Jensen, M. C., & Meckling, W. H. (1999). Specific Knowledge and Divisional Performance Measurement. *Journal of Applied Corporate Finance, 12*(2), 8-17.

Jensen, M. C., & Murphy, K. (1990). Performance Pay and Top-Management Incentives. *Journal of Political Economy, 98*(2), 225-262.

Kaplan, R., & Norton, D. (1996). Using the Balanced Scorecard as a Strategic Management System. *Harvard Business Review, 85*(7/8), 150-161.

Kay, J. A. (1976). Accountants, too, could be happy in a golden age: The accountant's rate of profit and the internal rate of return. *Oxford Economic Papers, 28*, 447-460.

Kim, J. J., Ahn, J.-H., & Yun, J. K. (2004, Spring). Economic Value Added (EVA) as a Proxy for Market Value Added (MVA) and Accounting Earnings: Empirical Evidence from the Business Cycle. *Journal of Accounting and Finance Research*, 40-48.

Kleiman, R. (1999). Some New Evidence on EVA Companies. *Journal of Applied Corporate Finance, 12*, 80-91.

Knight, J. A. (1998). *Value Based Management: Developing a Systematic Approach to Creating Shareholder Value*. New York: McGraw-Hill.

Kolk, A., van der Veen, M., Pinkse, J., & Fontanier, F. (2005). *KPMG International Survey of Corporate Responsability Reporting*. Amsterdam: KPMG Global Sustainability Services.

Koller, T. (1994). What is value-based management. *The McKinsey Quarterly, 3*, 87-101.

KPMG. (2003). *Valuation Snapshot: Wertorientierte Unternehmensführung. E_RIC - Eine Alternative Shareholder-Value-Spitzenkennzahl*. München: KPMG Deutsche Treuhand-Gesellschaft.

Kramer, J. K., & Peters, J. R. (2001). An Interindustry Analysis of Economic Value Added as a Proxy for Market Value Added. *Journal of Applied Finance, 11*(1), 41-49.

Kramer, J. K., & Pushner, G. (1997). An Empirical Analysis of Economic Value Added as a Proxy for Market Value Added. *Financial Practice and Education, 7*(1), 41-49.

Krotter, S. (2007). Zur Relevanz von Kongruenzdurchbrechungen für die Bewertung und Performance-Messung mit Residualgewinnen. *DBW, 6*, 692-718.

Kunz, A. H., Pfeiffer, T., & Schneider, G. (2007). E_RIC^{TM} versus EVA^{TM}. *Die Betriebswirtschaft, 67*(3), 259-277.

L.E.K. Consulting. (1998a). *Shareholder Value Added (Vol. I): Identifying and Managing Key Value Drivers* (Vol. I). Boston: L.E.K. Consulting.

L.E.K. Consulting. (1998b). *Shareholder Value Added (Vol. II): Creating an Ownership-Oriented Culture* (Vol. II). Boston: L.E.K. Consulting.

Lee, C., Myers, J., & Swaminathan, B. (1999). What is the Intrinsic Value Value of the Dow? *Journal of Finance, 54*, 1693-1741.

Lev, B. (1989). On the Usefulness of Earnings and Earnings Research: Lessons and Directions from Two Decades of Empirical Research. *Journal of Accounting Research, 27*(3), 153-192.

Lipe, R. C. (1986). The Information Contained in the Components of Earnings. *Journal of Accounting Research, 24*(3), 37-64.

Lynch, R. L., & Cross, K. F. (1995). *Measure Up!* Cambridge, MA: Blackwell Publishers.

Madden, B. J. (1998). The CFROI Valuation Model. *The Journal of Investing, Spring*, 31-44.

Madden, B. J. (1999). *CFROI Valuation: A Total System Approach to Valuing the Firm*. Oxford: Butterworth & Heinemann.

Marked by the market. (2001, December). *The Economist, 361*, 59-62.

Marshall, A. (1890). *Principles of Economics*. London, New York: The MacMillan Press Ltd.

Martin, J. D., & Petty, J. W. (2000). *Value based management: the corporate response to the shareholder revolution*. Boston, Massachusetts: Harvard Business School Press.

Mc Taggart, J., Kontes, P., & Mankins, M. (1994). *The value imperative*. New York: Free Press.

McCormack, J. L., & Vytheeswaran, J. (1998). How to Use EVA in the Oil and Gas Industry. *Journal of Applied Corporate Finance, 11*(3), 109-131.

Mergerstat. (1993). *Mergerstat Review*. Santa Monica: FactSet - Global Mergers and Acquisitions Information.

Milbourn, T. T. (1996). The Executive Compensation Puzzle: Theory and Evidence. London Business School.

Miller, M. H. (1977). Debt and Taxes. *Journal of Finance, 32*(2), 261-275.

Millman, G. (1997). Taking the Lies out of Earnings. *Worth, February*.

Milunovich, S., & Tsuei, A. (1996). EVA in the Computer Industry. *Journal of Applied Corporate Finance, 9*(1), 104-115.

Mitchell, R. K., Agle, B. R., & Wood, D. J. (1997). Toward a Theory of Stakeholder Identification and Salience: Defining the Principle of Who and What Really Counts. *Academy of Management Review, 22*(4), 853–886.

Modigliani, F., & Miller, M. H. (1958). The Cost of Capital, Corporation Finance and the Theory of Investment. *American Economic Review 48*(3), 261-297.

O'Byrne, S. F. (1996). EVA and market value. *Journal of Applied Corporate Finance, 12*(2), 116-125.

O'Byrne, S. F. (2000). Does Value Based Management Discourage Investment in Intangibles. In F. J. Fabozzi & J. L. Grant (Eds.), *Value-Based Metrics: Foundations and Practice* (pp. 99-132). New Hope, Pennsylvania: Frank J. Fabbozzi Associates.

O'Byrne, S. F., & Young, S. D. (2001). *EVA® and Value-Based Management: A Practical Guide to Implementation*. New York: McGraw-Hill.

O'Byrne, S. F., & Young, S. D. (2006). Incentives and Investor Expectations. *Journal of Applied Corporate Finance, 18*(2), 98-105.

O'Hanlon, J., & Peasnell, K. (2004). Residual Income Valuation: Are Inflation Adjustments Necessary? *Review of Accounting Studies, 9*, 375–398.

O'Keefe, T. B., & Lundholm, R. J. (2001). On comparing residual income and discounted cash flow models for use in equity valuation: A response to Penman *Contemporary Accounting Research 18*(4), 693-696.

Obrycki, D. J., & Resendes, R. (2000). Economic Margin: The Link between EVA and CFROI. In F. J. Fabozzi & J. L. Grant (Eds.), *Value-Based Metrics: Foundations and Practice* (pp. 157-178). New Hope, Pennsylvania: Frank J. Fabbozzi Associates.

Ohlson, J. A. (1990). A Synthesis of Security Valuation Theory and the Role of Dividends, Cash Flows, and Earnings. *Contemporary Accounting Research, 6*, 648-676.

Ohlson, J. A. (1995). Earnings, Book Value, and Dividends in Security Valuation. *Contemporary Accounting Research, 11*(2), 661-687.

Olsen, E. E. (1996). Economic Value Added. *Perspectives*.

Olsen, E. E. (2003). Rethinking Value-Based Management. *Handbook of Business Strategy, 4*(1), 286-301.

Ottosson, E., & Weissenrieder, F. (1996). *CVA. Cash Value Added – a new method for measuring financial performance.*Unpublished manuscript, Gothenburg University - Studies in Financial Economics (no. 1).

Pacioli, F. F. L. (1494). *Summa de Arithmetica*. Venice.

Patell, J. M. (1976). Corporate Forecasts of Earnings per Share and Stock Price Behavior: Empirical Tests. *Journal of Accounting Research, 14*(2), 246-276.

Peasnell, K. V. (1981). On capital budgeting and income measurement. *ABACUS, 17*, 52-67.

Peasnell, K. V. (1982). Some formal connections between economic values and yields and accounting numbers. *Journal of Business Finance and Accounting, 9*, 361-381.

Penman, S. H. (1998). A Synthesis of Equity Valuation Techniques and the Terminal Value Calculation for the Dividend Discount Model. *Review of Accounting Studies, 2*, 303-323.

Penman, S. H., & Sougiannis, T. (1998). A Comparison of Dividend, Cash Flow, and Earnings Approaches to Equity Valuation. *Contemporary Accounting Research, 15*(3), 343-383.

Peterson, P. P. (2000). Value-Based Measures of Performance. In F. J. Fabozzi & J. L. Grant (Eds.), *Value-Based Metrics: Foundations and Practice* (pp. 67-97). New Hope, Pennsylvania: Frank J. Fabbozzi Associates.

Peterson, P. P., & Peterson, D. R. (1996). *Comparison of Alternative Performance Measures*. Charlottesville, Virginia: Research Foundation of The Institute of Chartered Financial Analysts.

Pettit, J. M., Desautel, E., Millar, D., Ahmad, A., & Singer, J. (2000). Best of Times, Worst of Times. *EVAluation (by Stern Stewart & Co.), November.*

Pettit, J. M., & Goldberg, M. (2000). The New Math: 4 > 8. *EVAluation (by Stern Stewart & Co.), September.*

Pohlen, T. L., & Coleman, B. J. (2005). Evaluating Internal Operations and Supply Chain Performance Using EVA and ABC. *SAM Advanced Management Journal, Spring,* 45-58.

Prakash, A. J., Chang, C.-H., Davidson, L., & Lee, C.-C. (2003a). Adoption of Economic Value Added and Financial Ratios. *The International Journal of Finance, 15*(2), 2574-2592.

Prakash, A. J., Chang, C.-H., Davidson, L., & Lee, C.-C. (2003b). Adoption of Economic Value Added and Firm Risk. *The International Journal of Finance, 15*(2), 2593-2603.

Preinreich, G. A. D. (1936). The law of goodwill. *The Accounting Review, 12,* 317-329.

Preinreich, G. A. D. (1937). Goodwill in accountancy. *The Journal of Accountancy, July,* 28-50.

Preinreich, G. A. D. (1938). Annual Survey of Economic Theory: The Theory of Depreciation. *Econometrica, 6*(3), 219-241.

Ramana, D. V. (2007). Economic Value Added and Other Accounting Performance Indicators: An Empirical Analysis of Indian Companies. *The Icfai Journal of Accounting Research, VI*(2), 7-20.

Rappaport, A. (1986). *Creating Shareholder Value. The New Standard for Business Performance.* New York: The Free Press.

Rappaport, A. (1998). *Creating Shareholder Value: A Guide for Managers and Investors.* Southampton: B&T.

Rappaport, A. (2001). Scoreboard's Originator Answers the Questions Investors Ask the Most. *Wall Street Journal - Eastern Edition, 237*(39), R4.

Rayburn, J. (1986). The Association of Operating Cash Flow and Accruals with Security Returns *Journal of Accounting Research, 24*(3), 112-133.

Rice, V. (1996). Why EVA Works for Varity. *Chief Executive, 110,* 40-41.

Riceman, S. S., Cahan, S. F., & Lal, M. (2002). Do managers perform better under EVA bonus schemes? *The European Accounting Review, 11*(3), 537-572.

Ritter, J. R. (1991). The Long-Run Performance of Initial Public Offerings. *The Journal of Finance, XLVI*(1), 3-27.

Ritter, J. R., & Loughran, T. (1995). The New Issues Puzzle. *The Journal of Finance, L*(1), 23-51.

Ritter, J. R., & Warr, R. S. (2002). The Decline of Inflation and the Bull Market of 1982-1999. *Journal of Financial and Quantitative Analysis, 37*(1), 29-61.

Ross, I. (1995). The 1994 Stern Stewart Performance 1000. *Journal of Applied Corporate Finance, 7*(4), 105-110.

Ross, I. (1996). The Stern Stewart Performance 1000. *Journal of Applied Corporate Finance, 8*(4), 107-111.

Ross, I. (1997). The 1996 Stern Stewart Performance 1000. *Journal of Applied Corporate Finance, 9*(4), 115-120.

Ross, I. (1998). The Stern Stewart Performance 1000. *Journal of Applied Corporate Finance, 10*(4), 116-120.

Ross, I. (1999). The Stern Stewart Performance 1000. *Journal of Applied Corporate Finance, 11*(4), 122-126.

Ryan, H. E., & Trahan, E. A. (1999). The Utilization of Value-Based Management: An Empirical Analysis. *Financial Practice and Education, Spring/Summer*, 46-58.

Ryan, H. E., & Trahan, E. A. (2007). Corporate Financial Control Mechanisms and Firm Performance: The Case of Value-Based Management Systems. *Journal of Business Finance & Accounting, 34*(1&2), 111-138.

Sandoval, E. (2001). Financial Performance Measures and Shareholder Value Creation: An Empirical Study For Chilean Companies. *The Journal of Applied Business Research, 17*(3), 109-122.

Schremper, R., & Pälchen, O. (2001). Wertrelevanz rechnungswesenbasierter Erfolgskennzahlen - Eine empirische Untersuchung anhand des S&P 400 Industrial. *Die Betriebswirtschaft, 61*(5), 542-559.

Schüler, A. (1998). *Performance-Messung und Eigentümerorientierung*. Frankfurt am Main: Peter Lang.

Shinder, M., & McDowell, D. (1999). ABC, The Balanced Scorecard and EVA. *EVAluation (by Stern Stewart & Co.), 1*(2), 1-5.

Shrieves, R. E., & Wachowicz Jr., J. M. (2001). Free cash flow (FCF), economic value added (EVA), and net present value (NPV): A reconciliation of variations of discounted-cash-flow (DCF) valuation. *The Engineering Economist, 46*(1), 33-52.

Singh, K. P., & Garg, M. C. (2004). *Economic value added (EVA) in Indian corporates*. New Delhi: Deep & Deep.

Sirower, M. L., & O'Byrne, S. F. (1998). The Measurement of Post-Acquisition Performance: Toward a Value-Based Benchmarking Methodology. *Journal of Applied Corporate Finance, 11*(2), 107-121.

Smith, A. (1776). *An Inquiry into the Nature and Causes of the Wealth of Nations*.

Stelter, D. (1999). Wertorientierte Anreizsysteme. In W. a. T. S. Bühler (Ed.), *Wertorientierte Anreizsysteme für Führungskräfte und Manager* (pp. 207-241). Stuttgart.

Stelter, D., Riedl, J. B., & Plaschke, F. J. (2001). Wertschaffungskennzahlen und Bewertungsverfahren (1.4.3). In Achleitner/Thomas (Ed.), *Handbuch Corporate Finance* (2nd Ed., Loseblattausgabe ed.). Köln: Deutscher Wirtschaftsdienst.

Stern, E. (2000, September 24, 2000). Go for Real Growth. *The Sunday Times*,

Stern, J. M. (1980). *Analytical Methods in Financial Planning*. New York: Stern Stewart & Company.

Stern, J. M. (1990). One way to build value in your firm, a la executive compensation. *Financial Executive, November/December*, 51-54.

Stern, J. M. (1994). EVA Roundtable. *Journal of Applied Corporate Finance, 7*(2), 46-70.

Stewart, G. B. (1990). Announcing the Stern Stewart Performance 1,000: A New Way of Viewing Corporate America. *Journal of Applied Corporate Finance, 3*(2), 38-55.

Stewart, G. B. (1991). *The quest for value : the EVA TM management guide.* New York: HarperBusiness.

Stewart, G. B. (1994). EVA – fact or fantasy. *Journal of Applied Corporate Finance, 7*(2), 71-84.

Stewart, G. B. (2002). Accounting is Broken. Here's How to Fix It. A Radical Manifesto. *EVAluation (by Stern Stewart & Co.), 5*(1).

Stewart, G. B. (2003). How to Fix Accounting - Measure and Report Economic Profit. *Journal of Applied Corporate Finance, 15*(3), 63-82.

Stewart, T. A. (2005). Asking the Right Questions. *Harvard Business Review, 83*, 10.

Straw, R., Peck, S., & Keller, H.-U. (2000). EVA and the OECD Principles of Corporate Governance. In F. J. Fabozzi & J. L. Grant (Eds.), *Value-Based Metrics: Foundations and Practice* (pp. 195-227). New Hope, Pennsylvania: Frank J. Fabbozzi Associates.

Strong, N., & Walker, M. (1993, April). The Explanatory Power of Earnings for Stock Returns. *The Accounting Review, 68*(2), 385-399.

Tham, J. (2001a). Consistent Value Estimates from the Discounted Cash Flow (DCF) and Residual Income (RI) Models in M&M worlds without and with taxes [Electronic Version]. *Social Science Research Network Electronic Paper Collection, October,* from http://www.ssrn.com/

Tham, J. (2001b). Equivalence between Discounted Cash Flow (DCF) and Residual Income (RI) [Electronic Version]. *Social Science Research Network Electronic Paper Collection, February,* from http://www.ssrn.com/

Tham, J. (2001c). The Unbearable Lightness of EVA® in Valuation [Electronic Version]. *Social Science Research Network Electronic Paper Collection, April,* from http://www.ssrn.com/

Tham, J. (2004). EVA® Made simple: Is it Possible? [Electronic Version]. *Social Science Research Network Electronic Paper Collection, February,* from http://www.ssrn.com/

The Conference Board. (1999). Financial Assets & Equity Holdings. *Institutional Investment Report, 3.*

The Conference Board. (2000). Turnover, Investment Strategies, and Ownership Patterns. *Institutional Investment Report, 3.*

Tsuji, C. (2006). Does EVA beat earnings and cash flow in Japan? *Applied Financial Economics, 16*, 1199-1216.

Tully, S. (1993). The real key to creating wealth. *Fortune 128*(16), 38-50.

Tully, S. (1998). America's Greatest Wealth Creators. *Fortune 138*(9), 193-204.

Turvey, C. G., Lake, L., van Duren, E., & Sparling, D. (2000). The Relationship Between Economic Value Added and the Stock Market Performance of Agribusiness Firms. *Agribusiness, 16*(4), 399-416.

Uyemura, D. G., Kantor, C. C., & Pettit, J. M. (1996). EVA for Banks: Value Creation, Risk Management, and Profitability Measurement. *Journal of Applied Corporate Finance, 9*(2), 94-113.

v. Milano, G. (2000). EVA and the "New Economy". *Journal of Applied Corporate Finance, 13*(2), 118-128.

Vélez-Pareja, I. (2001). Economic Value Measurement: Investment Recovery and Value Added - IRVA [Electronic Version]. *Social Science Research Network Electronic Paper Collection, February,* from http://www.ssrn.com/

Velthius, L. J. (2004). *Value Based Management auf Basis von E_RIC.*Unpublished manuscript, Frankfurt am Main.

von Hayek, F. (1969). *The Corporation in a Democratic Society: In Whose Interest Ought It and Will It Be Run.* London: Penguin.

Waddock, S., & Graves, S. (2000). Performance Characteristics of Social and Traditional Investments. *Journal of Investing, 9*(2), 27-38.

Walbert, L. (1994). The Stern Stewart Performance 1000: Using EVA™ to Build Market Value. *Journal of Applied Corporate Finance, 6*(4), 109-112.

Walbert, L. (1995). The 1994 Stern Stewart Performance 1000. *Journal of Applied Corporate Finance, 7*(4), 105-110.

Wallace, J. S. (1997). Adopting residual income-based compensation plans: Do you get what you pay for? *Journal of Accounting and Economics, 24,* 275-300.

Wallace, J. S. (2003). Value Maximization and Stakeholder Theory: Compatible or Not. *Journal of Applied Corporate Finance, 15*(3), 120-127.

Warr, R. S. (2005). An empirical study of inflation distortions to EVA. *Journal of Economics and Business, 57,* 119-137.

Weaver, S. C. (2001). Measuring Economic Value Added: A Survey of the Practices of EVA Proponents. *Journal of Applied Finance, 11*(1), 50-60.

Weissenrieder, F. (1997). *Value Based Management: Economic Value Added or Cash Value Added?* .Unpublished manuscript, Gothenburg University - Studies in Financial Economics (no. 3).

West, A., & Worthington, A. C. (2004). Australian Evidence Concerning the Information Content of Economic Value-Added. *Australian Journal of Management, 29*(2), 201-224.

White, H. (1980). A heteroskedasticity-consistent covariance matrix estimator and a direct test for heteroskedasticity. *Econometrica, 48,* 817–838.

Wilson, G. P. (1986). The Relative Information Content of Accruals and Cash Flows: Combined Evidence at the Earnings Announcement and Annual Report Release Date. *Journal of Accounting Research, 24*(3), 165-200.

Wilson, G. P. (1987). The Incremental Information Content of the Accrual and Funds Components of Earnings After Controlling for Earnings. *The Accounting Review, 62*(2), 293-322.

Wilson, M. (2003). Corporate sustainability: What is it and where does it come from? *IVEY Business Journal, March/April,* 1-5.

Wolin, J. L., & Klopukh, S. (2000). Integrating EVA into the Portfolio Management Process. In F. J. Fabozzi & J. L. Grant (Eds.), *Value-Based Metrics: Foundations and Practice* (pp. 133-156). New Hope, Pennsylvania: Frank J. Fabbozzi Associates.

Yook, K. C. (1999). Estimating EVA Using Compustat PC Plus. *Financial Practice and Education, 9*(2), 33-37.

Yook, K. C. (2004). The Measurement of Post-Acquisition Performance Using EVA. *Quarterly Journal of Business & Economics, 42*(3/4), 67-83.

Young, S. D. (1999). Some Reflections on Accounting Adjustments and Economic Value Added. *Journal of Financial Statement Analysis, 4*(2), 7-19.

Appendix

Component	Source
Net income before extraordinary items / preferred dividends	WC01551
Add: Implied financing expenses	
AT interest expense, net of capitalized interest	WC01251, WC01255
Minority interest expense	WC01501
AT interest on average PV of non-capitalized leases @ 10%	Annual report
Subtract: Non-operating income (loss)	
AT interest income	WC01266
Other non-operating income (loss) *	See footnote
Subtract: Additional compensatory expenses	
Net pension and postretirement expense less service cost	Annual report
Cash contribution to ESOP	Annual report
Fair value reserve of stock options, net of tax	Annual report
Add: Increases in equity equivalents **	
Increase in deferred taxes	WC03263
Increases in LIFO, valuation, bad debt and other allowances	Annual report
Increase in net capitalized R&D expenses over 5 years	WC01201
Increase in net capitalized advertising expenses over 2 years	Annual report
AT goodwill amortization	Annual report
AT extraordinary credit (charge)	WC01253, WC01254

* Item defined as plug figure: Operating income, WC01250, interest income, WC01266, capitalized interest, WC01255, and extraordinary credit, WC01253, less extraordinary charge, WC01254, gross interest expense, WC01251, and pretax income, WC01401, times 1 minus the effective tax rate and pretax income, WC01401, less income tax expense, WC01451, minority interest expense, WC01501, and net earnings, WC01551.

** Equity equivalents defined according to Stewart (1991, pp. 112-117). Non-capitalized leases classified as debt equivalent.

Table 34   Used NOPAT Calculation Methodology

Component	Source
Average net assets	WC03999
Subtract: Average CIP and other investments	Annual report, WC02250
Subtract: Non-interest bearing current liabilities (NIBCLS) *	WC03101, WC03051
Add: Average equity and debt equivalents	
LIFO, valuation, bad debt and other allowances	Annual report
Net capitalized intangibles, see R&D and advertising expenses	Annual report
Cumulative AT goodwill amortization, starting 1994	Annual report
Cumulative AT unusual loss, since 1991	WC01254, WC01253, WC01601
Average PV of non-capitalized operating leases @ 10%	Annual report
Average net pension and postretirement liabilities	Annual report

*   Accounts payables and current accrued liabilities; defined via current liabilities, WC03101, less short-term debt & current portion of long-term debt, WC03051.

Table 35   Used Capital Calculation Methodology

Name	Sector	ISIN	Status
4kids Entertainment	Leisure Goods	US3508651011	Active
7-Eleven	Food & Drug Retailers	US8178262098	Suspended
99 Cents Only Stores	General Retailers	US65440K1060	Active
Advo	Media	US0075851024	Active
Airtran Holdings	Travel & Leisure	US00949P1084	Active
Alaska Air Group	Travel & Leisure	US0116591092	Active
Albertsons	Food & Drug Retailers	US0131041040	Dead
American Greetings 'A'	Media	US0263751051	Active
AMR (American Airlines)	Travel & Leisure	US0017651060	Active
Anheuser-Busch Companies	Beverages	US0352291035	Active
Ann Taylor Stores	General Retailers	US0361151030	Active
Applica	Household Goods	US03815A1060	Active
Aptargroup	General Industrials	US0383361039	Active
Arbitron	Media	US03875Q1085	Active
Archer-Danls.-Midl.	Food Producers	US0394831020	Active
Autonation	General Retailers	US05329W1027	Active
Avon Products	Personal Goods	US0543031027	Active
Aztar	Travel & Leisure	US0548021031	Dead
Bally Total Fitness Holding	Travel & Leisure	US05873K1088	Active
Best Buy	General Retailers	US0865161014	Active
Bestfoods	Food Producers	US08658U1016	Dead
Big Lots	General Retailers	US0893021032	Active
Bindley Wstn.Industries	Food & Drug Retailers	US0903241042	Dead
BJ's Wholesale Club	General Retailers	US05548J1060	Active
Borders Group	General Retailers	US0997091071	Active
Boyd Gaming	Travel & Leisure	US1033041013	Active
Briggs and Stratton	Household Goods	US1090431099	Active
Brinker International	Travel & Leisure	US1096411004	Active
Brown-Forman 'B'	Beverages	US1156372096	Active
Brunswick	Leisure Goods	US1170431092	Active
Cablevision Systems	Media	US12686C1099	Active
Callaway Golf	Leisure Goods	US1311931042	Active
Campbell Soup	Food Producers	US1344291091	Active
Carnival	Travel & Leisure	PA1436583006	Active
Carter Wallace	Household Goods	US1462851015	Dead
CEC Entertainments	Travel & Leisure	US1251371092	Active
Centex	Household Goods	US1523121044	Active
Champion Enterprises	Household Goods	US1584961098	Active
Chemfab	Personal Goods	US16361L1026	Dead
Chiquita Brands International	Food Producers	US1700328099	Active
CKE Restaurants	Travel & Leisure	US12561E1055	Active

Clear Channel Communications	Media	US1845021021	Active
Clorox	Household Goods	US1890541097	Active
Coachmen Industries	Leisure Goods	US1898731021	Active
Coca Cola	Beverages	US1912161007	Active
Cole National 'A'	General Retailers	US1932901036	Dead
Colgate-Palm.	Personal Goods	US1941621039	Active
Conagra Foods	Food Producers	US2058871029	Active
CSS Industries	Media	US1259061075	Active
CSX	Industrial Transportation	US1264081035	Active
Culp	Personal Goods	US2302151053	Active
Darden Restaurants	Travel & Leisure	US2371941053	Active
Dean Foods	Food Producers	US2423611035	Dead
Devry	General Retailers	US2518931033	Active
Dole Food	Food Producers	US2566051064	Dead
Dollar General	General Retailers	US2566691026	Active
Entravision Communications 'A'	Media	US29382R1077	Active
Ethan Allen Interiors	Household Goods	US2976021046	Active
Extended Stay American	Travel & Leisure	US30224P1012	Dead
Family Dollar Stores	General Retailers	US3070001090	Active
Federated Department Stores	General Retailers	US31410H1014	Active
Fedex	Industrial Transportation	US31428X1063	Active
Fleetwood Enterprises	Leisure Goods	US3390991038	Active
Florida East Coast Industries	Industrial Transportation	US3406321089	Active
Fort James	Leisure Goods	US3474711047	Dead
Furniture Brands International	Household Goods	US3609211004	Active
Gannett	Media	US3647301015	Active
GAP	General Retailers	US3647601083	Active
Gaylord Entertainment	Travel & Leisure	US3679051066	Active
General Mills	Food Producers	US3703341046	Active
Gillette	Personal Goods	US3757661026	Dead
GT.Atl.and Pacific	Food & Drug Retailers	US3900641032	Active
Gtech Holdings	Travel & Leisure	US4005181064	Dead
Guest Supply	Personal Goods	US4016301081	Dead
H and R Block	General Retailers	US0936711052	Active
Hancock Fabrics	General Retailers	US4099001079	Active
Handleman	Media	US4102521006	Active
Hannaford Brothers	Food & Drug Retailers	US4105501070	Dead
Harman International Industries	Leisure Goods	US4130861093	Active
Harsco	General Industrials	US4158641070	Active
Harte-Hanks	Media	US4161961036	Active
Hartmarx	Personal Goods	US4171191046	Active
Hasbro	Leisure Goods	US4180561072	Active

Haverty Furniture Companies	General Retailers	US4195961010	Active
Hilton Hotels	Travel & Leisure	US4328481092	Active
HNI	Household Goods	US4042511000	Active
Home Depot	General Retailers	US4370761029	Active
Honeywell International	General Industrials	US4385161066	Active
Hormel Foods	Food Producers	US4404521001	Active
IBP	Food Producers	US4492231067	Dead
International Game Technology	Travel & Leisure	US4599021023	Active
International Multifoods	Food Producers	US4600431021	Dead
Interpool	Industrial Transportation	US46062R1086	Active
Jack In The Box	Travel & Leisure	US4663671091	Active
Jones Apparel Group	Personal Goods	US4800741039	Active
KB Home	Household Goods	US48666K1097	Active
Kimberly-Clark	Personal Goods	US4943681035	Active
Kirby	Industrial Transportation	US4972661064	Active
Knight Transportation	Industrial Transportation	US4990641031	Active
Kohls	General Retailers	US5002551043	Active
Landrys Restaurants	Travel & Leisure	US51508L1035	Active
LA-Z-Boy Chair	Household Goods	US5053361078	Active
LEE Enterprises	Media	US5237681094	Active
Limited Brands	General Retailers	US5327161072	Active
Lithia Motors A	General Retailers	US5367971034	Active
Lowe's Companies	General Retailers	US5486611073	Active
Luby	Travel & Leisure	US5492821013	Active
Martha Stewart Living Omnimedia 'A'	Media	US5730831022	Active
Marvel Entertainment	Leisure Goods	US57383T1034	Active
Mattel	Leisure Goods	US5770811025	Active
May Department Stores	General Retailers	US5777781031	Dead
Mccormick and Co Non-Voting	Food Producers	US5797802064	Active
Meadwestvaco	General Industrials	US5833341077	Active
Mens Wearhouse	General Retailers	US5871181005	Active
Meredith	Media	US5894331017	Active
MGM Mirage	Travel & Leisure	US5529531015	Active
Michaels Stores	General Retailers	US5940871081	Dead
Mirage Resorts	Travel & Leisure	US60462E1047	Dead
ML Macadamia Orchards LP DEP Unit 'A'	Food Producers	US55307U1079	Active
Mohawk Industries	Household Goods	US6081901042	Active
Myers Industries	General Industrials	US6284641098	Active
National Presto Industries	Household Goods	US6372151042	Active
Neiman-Marcus Group 'A'	General Retailers	US6402042021	Dead
Nelson Thomas	Media	US6403761090	Suspended
New England Business Service	General Retailers	US6438721044	Dead

New York Times 'A'	Media	US6501111073	Active
Newell Rubbermaid	Household Goods	US6512291062	Active
Nike 'B'	Personal Goods	US6541061031	Active
Nordstrom	General Retailers	US6556641008	Active
Norfolk Southern	Industrial Transportation	US6558441084	Active
Nortek Holdings	Household Goods	US6565571055	Dead
Office Depot	General Retailers	US6762201068	Active
Oil-DRI American	Household Goods	US6778641000	Active
Omnicare	Food & Drug Retailers	US6819041087	Active
Oxford Industries	Personal Goods	US6914973093	Active
Penney JC	General Retailers	US7081601061	Active
Pepsico	Beverages	US7134481081	Active
Phosphate Resource Partners Depositary Unit	Food Producers	US7192171012	Suspended
Pier 1 Imports	General Retailers	US7202791080	Active
Playtex Products	Personal Goods	US72813P1003	Active
Polaris Industries	Leisure Goods	US7310681025	Active
PRE Paid Legal Services	General Retailers	US7400651078	Active
Procter and Gamble	Household Goods	US7427181091	Active
Quaker Oats	Food Producers	US7474021059	Dead
Quiksilver	Personal Goods	US74838C1062	Active
Radioshack	General Retailers	US7504381036	Active
Readers Digest	Media	US7552671015	Active
Reebok International	Personal Goods	US7581101000	Dead
Rent Way	General Retailers	US76009U1043	Dead
Republic Group	Leisure Goods	US7604731080	Dead
Revlon 'A'	Personal Goods	US7615255004	Active
REX Stores	General Retailers	US7616241052	Active
Richfood Holdings A	Food & Drug Retailers	US7634081019	Dead
Rock Tenn 'A' Shares	General Industrials	US7727392075	Active
Rockwell Automation	General Industrials	US7739031091	Active
Rollins	General Retailers	US7757111049	Active
Rollins Truck Leasing	Industrial Transportation	US7757411019	Dead
Ruby Tuesday	Travel & Leisure	US7811821005	Active
Russell	Personal Goods	US7823521080	Dead
Sara LEE	Food Producers	US8031111037	Active
Servicemaster	General Retailers	US81760N1090	Active
Shopko Stores	General Retailers	US8249111019	Dead
Smithfield Foods	Food Producers	US8322481081	Active
Smucker JM	Food Producers	US8326964058	Active
Snap-on	Household Goods	US8330341012	Active
Sotheby's	General Retailers	US8358981079	Active
Southwest Airlines	Travel & Leisure	US8447411088	Active

Speedway Motorsports	Travel & Leisure	US8477881069	Active
Standard Commercial	Food & Drug Retailers	US8532581019	Suspended
Stanley Works	Household Goods	US8546161097	Active
Starrett LS	Household Goods	US8556681091	Active
Steinway Musical Instruments	Leisure Goods	US8584951045	Active
Stride Rite	Personal Goods	US8633141002	Active
Supervalu	Food & Drug Retailers	US8685361037	Active
Talbots	General Retailers	US8741611029	Active
Textron	General Industrials	US8832031012	Active
Thor Industries	Leisure Goods	US8851601018	Active
Timberland 'A'	Personal Goods	US8871001058	Active
Toll Brothers	Household Goods	US8894781033	Active
Tootsie Roll	Food Producers	US8905161076	Active
Triarc Companies 'A'	Travel & Leisure	US8959271011	Active
Tyson Foods 'A'	Food Producers	US9024941034	Active
Unifi	Personal Goods	US9046771013	Active
Unifirst	Personal Goods	US9047081040	Active
Union Pacific	Industrial Transportation	US9078181081	Active
United Parcel Service	Industrial Transportation	US9113121068	Active
United States Home New	Household Goods	US9119201062	Dead
V F	Personal Goods	US9182041080	Active
Valassis Communications	Media	US9188661048	Active
Walgreen	Food & Drug Retailers	US9314221097	Active
Wallace Computer Service	Media	US9322701015	Dead
Walter Industries	General Industrials	US93317Q1058	Active
Weis Markets	Food & Drug Retailers	US9488491047	Active
Wendy's International	Travel & Leisure	US9505901093	Active
West Pharmaceutical Services	General Industrials	US9553061055	Active
Williams Sonoma	General Retailers	US9699041011	Active
Winnebago Industries	Leisure Goods	US9746371007	Active
WMS Industries	Travel & Leisure	US9292971093	Active
Wolverine Worldwide	Personal Goods	US9780971035	Active
Wrigley William Junior	Food Producers	US9825261053	Active
Xtra	Industrial Transportation	US9841381075	Dead

Table 36   List of Firms Contained in the Sample

Rank	(1)	>	(2)	>	(3)	>	(4)	>	(5)	>	(6)
A: Equal coefficients											
Measure	RI		EBEI		CRI		CFO		EVA		CEVA
Adj. R^2	0.184		0.165		0.154		0.078		0.030		0.025
p-value		(0.038)		(0.000)		(0.000)		(0.000)		(1.000)	
				(1.000)		(0.000)		(0.002)		(0.000)	
						(0.000)		(0.001)		(0.109)	
								(0.005)		(0.023)	
										(0.049)	
B: Sign-based coefficients											
Measure	RI		CRI		EBEI		EVA		CEVA		CFO
Adj. R^2	0.270		0.260		0.252		0.193		0.174		0.091
p-value		(0.861)		(0.000)		(0.000)		(1.000)		(0.000)	
				(0.000)		(0.000)		(0.000)		(0.000)	
						(0.000)		(0.000)		(0.000)	
								(0.000)		(0.000)	
										(0.000)	

Table 37   Relative Information Content Tests from Associations with Firm Value – Including Outliers

Rank	(1)	>	(2)	>	(3)	>	(4)	>	(5)	>	(6)
A: Equal coefficients											
Measure	RI		EBEI		CRI		CFO		CEVA		EVA
Adj. R^2	0.156		0.145		0.128		0.125		0.025		0.019
p-value		(0.396)		(0.000)		(0.000)		(0.000)		(1.000)	
				(1.000)		(0.000)		(1.000)		(0.000)	
						(0.000)		(0.000)		(1.000)	
								(0.093)		(0.000)	
										(0.000)	
B: Sign-based coefficients											
Measure	EBEI		RI		CFO		CRI		EVA		CEVA
Adj. R^2	0.183		0.172		0.169		0.163		0.071		0.068
p-value		(0.000)		(0.000)		(0.000)		(0.999)		(0.915)	
				(0.000)		(0.000)		(0.000)		(0.995)	
						(0.000)		(0.993)		(0.000)	
								(0.291)		(0.566)	
										(0.000)	

Table 38   Relative Information Content Tests referring to Stock Returns – Including Outliers

Sequence		(1)	(2)	(3)	(4-CRI)	(4-CEVA)	(5-CEVA)
Component	CFO	Accr	ATInt	CapChg	DepAdj1	AccAdj	DepAdj2
A: Equal coefficients							
Δ Adj. R^2	0.078	0.095	0.034	0.000	0.006	0.001	0.008
p-value		(0.000)	(0.000)	(0.452)	(0.000)	(0.170)	(0.000)
B: Sign-based coefficients							
Δ Adj. R^2	0.091	0.099	0.028	0.000	0.004	0.015	0.006
p-value		(0.000)	(0.000)	(0.702)	(0.000)	(0.012)	(0.000)

Table 39   Incremental Information Content Tests referring to Firm Value – Including Outliers

Sequence		(1)	(2)	(3)	(4-CRI)	(4-CEVA)	(5-CEVA)
Component	CFO	Accr	ATInt	CapChg	DepAdj1	AccAdj	DepAdj2
A: Equal coefficients							
Δ Adj. R^2	0.125	0.082	0.001	0.037	0.001	-0.001	0.002
p-value		(0.042)	(0.985)	(0.009)	(0.629)	(0.958)	(0.550)
B: Sign-based coefficients							
Δ Adj. R^2	0.169	0.067	0.000	0.019	0.009	0.010	0.004
p-value		(0.001)	(0.982)	(0.264)	(0.157)	(0.085)	(0.300)

Table 40   Incremental Information Content Tests referring to Stock Returns– Including Outliers